Science and Technology of Chemical Mechanical Planarization (CMP)

MATERIALS RESEARCH SOCIETY
SYMPOSIUM PROCEEDINGS VOLUME 1157

Science and Technology of Chemical Mechanical Planarization (CMP)

Symposium held April 14–16, 2009, San Francisco, California, U.S.A.

EDITORS:

Ashok Kumar
University of South Florida
Tampa, Florida, U.S.A.

C. Fred Higgs III
Carnegie Mellon University
Pittsburgh, Pennsylvania, U.S.A.

Chad S. Korach
State University of New York-Stony Brook
Stony Brook, New York, U.S.A.

Subramanian Balakumar
National Center for Nanosciences
and Nanotechnology
Tamil Nadu, India

Materials Research Society
Warrendale, Pennsylvania

CAMBRIDGE
UNIVERSITY PRESS

University Printing House, Cambridge CB2 8BS, United Kingdom

One Liberty Plaza, 20th Floor, New York, NY 10006, USA

477 Williamstown Road, Port Melbourne, VIC 3207, Australia

314-321, 3rd Floor, Plot 3, Splendor Forum, Jasola District Centre, New Delhi - 110025, India

103 Penang Road, #05-06/07, Visioncrest Commercial, Singapore 238467

Cambridge University Press is part of the University of Cambridge.

It furthers the University's mission by disseminating knowledge in the pursuit of education, learning and research at the highest international levels of excellence.

www.cambridge.org
Information on this title: www.cambridge.org/9781605111308

Materials Research Society
506 Keystone Drive, Warrendale, PA 15086
http://www.mrs.org

© Materials Research Society 2010

First published 2010
First paperback edition 2012

Single article reprints from this publication are available through University Microfilms Inc., 300 North Zeeb Road, Ann Arbor, MI 48106

CODEN: MRSPDH

A catalogue record for this publication is available from the British Library

ISBN 978-1-605-11130-8 Hardback
ISBN 978-1-107-40830-2 Paperback

CONTENTS

*Invited Paper

v

CMP IN MEMORY AND
DATA STORAGE TECHNOLOGIES

TOOL/PROCESS DEVELOPMENT
SUCH AS eCMP AND LOW-SHEAR CMP

ADVANCED CMP PROCESS
CONTROL TECHNIQUES

*Invited Paper

PREFACE

Symposium E, "Science and Technology of Chemical Mechanical Planarization (CMP)," was held April 14–16 at the 2009 MRS Spring Meeting in San Francisco, California. This volume consists of invited, contributed, and poster presentations from national and international researchers representing universities, federal laboratories and industries. The objective of this symposium is to bring together experts from a broad spectrum of research and technology groups who are currently working on CMP challenges. This symposium addresses all aspects of integration challenges and key structural reliability issues for CMP of Cu-interconnects, MEMS, and compound semiconductors.

Ashok Kumar
C. Fred Higgs III
Chad S. Korach
Subramanian Balakumar

February 2010

MATERIALS RESEARCH SOCIETY SYMPOSIUM PROCEEDINGS

MATERIALS RESEARCH SOCIETY SYMPOSIUM PROCEEDINGS

Volume 1175E —Novel Functional Properties at Oxide-Oxide Interfaces, G. Rijnders, R. Pentcheva,
J. Chakhalian, I. Bozovic, 2009, ISBN 978-1-60511-148-3

Volume 1176E —Nanocrystalline Materials as Precursors for Complex Multifunctional Structures through
Chemical Transformations and Self Assembly, Y. Yin, Y. Sun, D. Talapin, H. Yang, 2009,
ISBN 978-1-60511-149-0

Volume 1177E —Computational Nanoscience — How to Exploit Synergy between Predictive Simulations
and Experiment, G. Galli, D. Johnson, M. Hybertsen, S. Shankar, 2009,
ISBN 978-1-60511-150-6

Volume 1178E —Semiconductor Nanowires — Growth, Size-Dependent Properties and Applications,
A. Javey, 2009, ISBN 978-1-60511-151-3

Volume 1179E —Material Systems and Processes for Three-Dimensional Micro- and Nanoscale Fabrication
and Lithography, S.M. Kuebler, V.T. Milam, 2009, ISBN 978-1-60511-152-0

Volume 1180E —Nanoscale Functionalization and New Discoveries in Modern Superconductivity,
R. Feenstra, D.C. Larbalestier, B. Maiorov, M. Putti, Y.-Y. Xie, 2009,
ISBN 978-1-60511-153-7

Volume 1181 — Ion Beams and Nano-Engineering, D. Ila, P.K. Chu, N. Kishimoto, J.K.N. Lindner,
J. Baglin, 2009, ISBN 978-1-60511-154-4

Volume 1182 — Materials for Nanophotonics — Plasmonics, Metamaterials and Light Localization,
M. Brongersma, L. Dal Negro, J.M. Fukumoto, L. Novotny, 2009,
ISBN 978-1-60511-155-1

Volume 1183 — Novel Materials and Devices for Spintronics, O.G. Heinonen, S. Sanvito, V.A. Dediu,
N. Rizzo, 2009, ISBN 978-1-60511-156-8

Volume 1184 — Electron Crystallography for Materials Research and Quantitative Characterization of
Nanostructured Materials, P. Moeck, S. Hovmöller, S. Nicolopoulos, S. Rouvimov,
V. Petkov, M. Gateshki, P. Fraundorf, 2009, ISBN 978-1-60511-157-5

Volume 1185 — Probing Mechanics at Nanoscale Dimensions, N. Tamura, A. Minor, C. Murray,
L. Friedman, 2009, ISBN 978-1-60511-158-2

Volume 1186E —Nanoscale Electromechanics and Piezoresponse Force Microcopy of Inorganic,
Macromolecular and Biological Systems, S.V. Kalinin, A.N. Morozovska, N. Valanoor,
W. Brownell, 2009, ISBN 978-1-60511-159-9

Volume 1187 — Structure-Property Relationships in Biomineralized and Biomimetic Composites,
D. Kisailus, L. Estroff, W. Landis, P. Zavattieri, H.S. Gupta, 2009,
ISBN 978-1-60511-160-5

Volume 1188 — Architectured Multifunctional Materials, Y. Brechet, J.D. Embury, P.R. Onck, 2009,
ISBN 978-1-60511-161-2

Volume 1189E —Synthesis of Bioinspired Hierarchical Soft and Hybrid Materials, S. Yang, F. Meldrum,
N. Kotov, C. Li, 2009, ISBN 978-1-60511-162-9

Volume 1190 — Active Polymers, K. Gall, T. Ikeda, P. Shastri, A. Lendlein, 2009,
ISBN 978-1-60511-163-6

Volume 1191 — Materials and Strategies for Lab-on-a-Chip — Biological Analysis, Cell-Material
Interfaces and Fluidic Assembly of Nanostructures, S. Murthy, H. Zeringue, S. Khan,
V. Ugaz, 2009, ISBN 978-1-60511-164-3

Volume 1192E —Materials and Devices for Flexible and Stretchable Electronics, S. Bauer, S.P. Lacour,
T. Li, T. Someya, 2009, ISBN 978-1-60511-165-0

Volume 1193 — Scientific Basis for Nuclear Waste Management XXXIII, B.E. Burakov, A.S. Aloy,
2009, ISBN 978-1-60511-166-7

Prior Materials Research Society Symposium Proceedings available by contacting Materials Research Society

Polishing, Conditioning, and
Wear Mechanisms on the Pad

Mater. Res. Soc. Symp. Proc. Vol. 1157 © 2009 Materials Research Society   1157-E01-01

Optimizing Pad Groove Design and Polishing Kinematics for Reduced Shear Force, Low Force Fluctuation and Optimum Removal Rate Attributes of Copper CMP

Yasa Sampurno [1,2], Ara Philipossian [1,2], Sian Theng [1,2], Takenao Nemoto [3], Xun Gu [3], Yun Zhuang [1,2], Akinobu Teramoto [3] and Tadahiro Ohmi [3]

[1] Araca, Inc., 2550 East River Road, Suite 12204, Tucson, Arizona 85718 USA
[2] University of Arizona, 1133 James E. Rogers Way, Tucson, Arizona 85721 USA
[3] Tohoku University, 6-6-10, Aza-Aoba, Aramaki, Aoba-ku, Sendai 90-8579 Japan

ABSTRACT

The effect of polisher kinematics on average and standard deviation of shear force and removal rate in copper CMP is investigated. A 'delamination factor' consisting of average shear force, standard deviation of shear force, and required polishing time is defined and calculated based on the summation of normalized values of the above three components. In general, low values of the 'delamination factor' are preferred since it is believed that they minimize defects during polishing. In the first part of this study, 200-mm blanket copper wafers are polished at constant platen rotation of 25 RPM and polishing pressure of 1.5 PSI with different wafer rotation rates and slurry flow rates. Results indicate that at the slurry flow rate of 200 ml/min, 'delamination factor' is lower by 14 to 54 percent than at 400 ml/min. Increasing wafer rotation rate from 23 to 148 RPM reduces 'delamination factor' by approximately 50 percent and improves removal rate within-wafer-non-uniformity by appx. 2X. In the second part of this study, polishing is performed at the optimal slurry flow rate of 200 ml/min and wafer rotation rate of 148 RPM with different polishing pressures and platen rotation rates. Results indicate that 'delamination factor' is reduced significantly at the higher ratio of wafer to platen rotation rates.

INTRODUCTION

During CMP, 2-body and 3-body interactions among the wafer, the slurry particles and the pad generate a wide range of shear forces. With IC devices shrinking further into nano-scale dimensions and low-k nano-porous materials becoming increasingly attractive, shearing needs to be re-visited vis-à-vis the mechanical integrity of the metal and underlying porous dielectric stack since high average shear force, in combination with its large fluctuating component and long polishing times (especially to clear the barrier) become major defect-causing problems. In this paper, the contributions of shear force, standard deviation of shear force and required polishing time are presented independently in a bar graph. The value of 'delamination factor' is calculated based the summation of normalized values of the above three components and normalized to illustrate the effect of polisher kinematics during copper CMP process.

EXPERIMENTAL APPARATUS AND PROCEDURE

All experiments were performed on an Araca APD-800 polisher and tribometer which is equipped with the unique ability to acquire shear force and down force in real-time.[1] In this study, force acquisition rate was set at 1,000 Hz. An IC1000 pad with Suba IV sub-pad was used. The pad was grooved with positive $20°$ slanted concentric grooves superimposed on $0°$

logarithmic positive grooves. A 100-grit triple ring dots (TRD) diamond disc from Mitsubishi Materials Corporation was used to condition the pad at 5.8 lb_f during wafer polishing. The conditioner disc rotated at 95 RPM and swept 10 times per minute. During polishing, the diamond disc, pad and wafer rotated counter-clockwise. Hitachi Chemical HS-2H-635-12 slurry was used and polishing time for each wafer was set to 90 seconds. In the first part of the study, the effect of slurry flow rate and wafer rotation rate was explored at constant platen rotation of 25 RPM and polishing pressure of 1.5 PSI. Slurry flow rate was set at 200, 300 or 400 ml/min. The wafer rotation rate was set at 23, 98 or 148 RPM. In the second part of the study, the effect of polishing pressure and platen rotation rate was explored using the optimal slurry flow rate (200 ml/min) and wafer rotation rate (148 RPM) obtained from the first part of the study. The polishing pressure was set at 1.5, 2 or 2.5 PSI. The platen rotation rate was varied at 25, 40 and 55 RPM.

THE 'DELAMINATION FACTOR'

During CMP, 2-body and 3-body interactions among the wafer, the slurry particles and the pad result in polished substrate removal and generate a wide range of shear forces. In general, higher average shear force and standard deviation of shear force, as well as longer polishing time contribute to delamination (and therefore defects) during polishing. In this paper, the contributions of average shear force, standard deviation of shear force and required polishing time are presented independently in a bar graph. The value of 'delamination factor' which is the summation of normalized values of the above three components represents this combined contribution. Figure 1 shows a 'delamination factor' graph of a reference actual polishing process (i.e. 5-min polishing with an average shear force of 12 lb_f and shear force standard deviation of 8 lb_f). For the reference process, the value of each component is unity in the bar graph of 'delamination factor'. Therefore, the value of 'delamination factor' is equal to 3.

If the example shown in Figure 1 is used as the reference process, Figure 2 shows the 'delamination factor' corresponding to a 2-minute polishing process with an average shear force of 7.2 lb_f and standard deviation of shear force of 4 lb_f. The normalized values are 0.6, 0.5 and 0.4 for the average shear force, standard deviation of shear force, and required polishing time, respectively. The 'delamination factor' value is therefore 1.5.

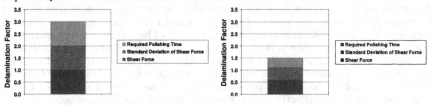

Figure 1. 'Delamination factor' of a reference process

Figure 2. 'Delamination factor' of an improved CMP process

ASYNCHRONOUS PAD AND WAFER ROTATION RATE

In a wafer-centered polar coordinate system, the pad-wafer relative velocity, V_{P-W}, at any point on the wafer (r,Θ) differs from the sliding velocity at the wafer center $(r=0)$ by a factor $\ddot{A}$

that depends on the wafer to platen rotation rate ratio Ω_w/Ω_p and the pad-wafer center separation R as follows: [2]

$$\tilde{A}(r,\theta) = \frac{v_{P-W}\,(r,\theta)}{v_{P-W}\,(0,\theta)} = \sqrt{1 + 2\left(1 - \frac{\Omega_W}{\Omega_P}\right)\left(\frac{r}{R}\right)\cos\theta + \left(1 - \frac{\Omega_W}{\Omega_P}\right)^2 \left(\frac{r}{R}\right)^2}$$ [1]

For synchronous pad-wafer rotation, the factor $\tilde{A}$ is unity at any point on the wafer (r,θ). For asynchronous pad-wafer rotation, the average asynchronous pad-wafer sliding velocity, A, across the wafer radius R_w can be calculated based on the area weighted distribution as follows:

$$A = \frac{\int_0^{2\pi}\int_0^{R_W} \tilde{A}(r,\theta)r\,dr\,d\theta}{\int_0^{2\pi}\int_0^{R_W} r\,dr\,d\theta}$$ [2]

Table 1 summarizes the average asynchronous pad-wafer sliding velocity, A, for five combinations of Ω_w and Ω_p used in this study. The value of A was calculated for 200-mm wafers with pad-wafer center separation, R, of 225 mm.

Ω_p	Ω_w	A
25	23	1.00
25	98	1.21
25	148	1.62
40	148	1.18
55	148	1.07

Table 1. The average asynchronous pad-wafer sliding velocity, A

RESULTS AND DISCUSSION

Effect of slurry flow rate and carrier rotation rate

In the first part of the study, the effect of slurry flow rate and wafer rotation rate is studied at a constant platen rotation of 25 RPM and polishing pressure of 1.5 PSI. This study aims to select the optimum slurry flow rate and wafer rotation rate to achieve lower 'delamination factor' values and better removal rate within-wafer-non-uniformity (RR WIWNU).

Figure 3(a) and 3(b) shows the average and standard deviation of shear force as a function of wafer rotation rate and slurry flow rate. Results indicate that increasing the wafer rotation rate from 23 RPM to 148 RPM reduces the average and standard deviation of shear force by approximately 40 percent and 50 percent, respectively. In comparison, there is no consistent trend in the effect of slurry flow rate on the average and standard deviation of shear force.

Figure 4(a) shows that higher wafer rotation rates lead to higher copper removal rates. As indicated in Table 1, at a constant platen rotation rate, higher wafer rotation rates result in higher sliding velocities, leading to higher removal rates. Figure 4(b) shows the normalized RR WIWNU as a function of wafer rotation rate and slurry flow rate. The RR WIWNU at wafer rotation rate of 23 RPM and slurry flow rate of 400 ml/min is employed as a baseline for normalization. Increasing the wafer rotation rate from 23 RPM to 148 RPM reduces the RR WIWNU by approximately 45 percent. At wafer and platen rotation rate of 23 RPM and 25 RPM, removal rate profile indicates a center-fast removal mechanism. Increasing the wafer rotation rate leads to higher local sliding velocity toward the wafer edge, enhances the local removal rate, and improves the RR WIWNU. In comparison, there is no consistent trend in the effect of slurry flow rate on the removal rate and RR WIWNU.

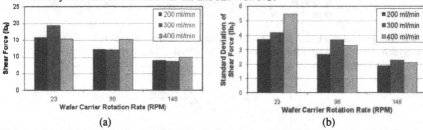

(a)   (b)

Figure 3. (a) Average shear force and (b) standard deviation of shear force as a function of wafer carrier rotation and slurry flow rate

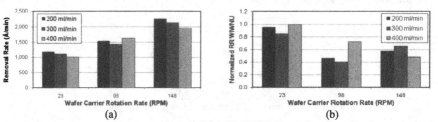

(a)   (b)

Figure 4. (a) Removal rate and (b) normalized RR WIWNU as a function of wafer carrier rotation and slurry flow rate

Figure 5 shows the 'delamination factor' values. The process at the wafer rotation rate of 23 RPM and slurry flow rate of 400 ml/min is used as the reference process for normalization. The results indicate that the 'delamination factor' values are 48 to 54 percent lower at the wafer rotation rate of 148 RPM compared with at 25 RPM. Figure 5 also indicates that the 'delamination factor' values increase with slurry flow rate. The lowest 'delamination factor' value is obtained when the slurry flow rate is at 200 ml/min and the wafer rotation rate at 148 RPM. Under this condition, the removal rate within-wafer-non-uniformity improves by a factor of appx. 2 compared to the reference process. Therefore, increasing the ratio of Ω_w/Ω_p leads to significantly lower values of 'delamination factor'.

Effect of polishing pressure and platen rotation rate

In the second part of the study, the effect of polishing pressure and platen rotation rate is explored under the optimal slurry flow rate (i.e. 200 ml/min) and wafer rotation rate (i.e. 148 RPM) obtained in the first part of the study. Table 2 summarizes the five different combinations of polishing parameters used in this part of the study. The No. 1 polishing parameter shown in Table 2 achieves the lowest 'delamination factor' reported in the first part of the study.

6

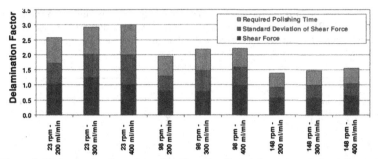

Figure 5. 'Delamination factor' as a function of wafer rotation rate and slurry flow rate

No.	Pressure (PSI)	Ω_p (RPM)	Ω_w (RPM)	Slurry Flow Rate (ml/min)
1	1.5	25	148	200
2	1.5	55	148	200
3	2	40	148	200
4	2.5	25	148	200
5	2.5	55	148	200

Table 2. Polishing conditions to explore the effect of polishing pressure and platen rotation rate

Figures 6(a) and (b) show a non-linear trend in the average and standard deviation of shear force as a function of the product of polishing pressure and sliding velocity (pV). At 2.5 PSI and platen rotation rate of 25 RPM, the average and standard deviation of shear force depart (i.e. decrease) significantly from the trend.

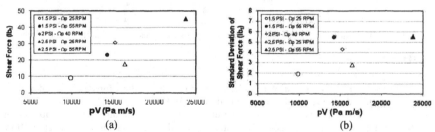

Figure 6. (a) Average shear force and (b) standard deviation of shear force as a function of pV

Figure 7(a) shows the removal rate as a function of the product of pV. The removal rate increases non-linearly with. However, unlike the trend of average and standard deviation of shear force, the removal rate at polishing pressure of 2.5 PSI and platen rotation rate of 25 RPM lies in the expected range. Figure 7(b) shows the normalized RR WIWNU using the same reference process as shown in Figure 4(b).

Figure 8 shows the 'delamination factor' values using the same reference process as shown in Figure 5. The 'delamination factor' value at the polishing pressure of 2.5 PSI and platen rotation rate of 25 RPM was close to that at the polishing pressure of 1.5 PSI and platen rotation rate of 25 RPM. Such a low 'delamination factor' value is due to its low shear force, low standard deviation of shear force coupled with its higher removal rate for a shorter required polishing time.

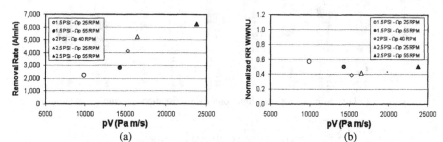

Figure 7. (a) Removal rate and (b) normalized RR WIWNU as a function of pV

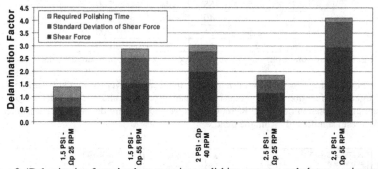

Figure 8. 'Delamination factor' values at various polishing pressure and platen rotation rate

SUMMARY

In this study, a 'delamination factor' consisting of average shear force, standard deviation of shear force, and required polishing time is defined, and its value is calculated based on the summation of normalized values of the above three components. Results show that optimizing polishing parameters, such as polishing pressure, wafer and platen rotation rates, and slurry flow rate, can significantly reduce the 'delamination' values and improve the removal rate within-wafer-non-uniformity.

REFERENCES

1. http://aracainc.com/products/apd-800
2. G. Muldowney, J. Hendron and T. Crkvenac, *Proceeding of the 10th International Conference on Chemical-Mechanical Polish (CMP) Planarization for ULSI Multilevel Interconnection (CMP-MIC)*, p. 168 (2005)

Mater. Res. Soc. Symp. Proc. Vol. 1157 © 2009 Materials Research Society   1157-E01-02

Pad Topography, Contact Area and Hydrodynamic Lubrication in Chemical-Mechanical Polishing

Leonard J. Borucki[1], Ting Sun[2], Yun Zhuang[1,2], David Slutz[3] and Ara Philipossian[1,2]

[1]Araca Incorporated, 6655 N. Canyon Crest Dr., Suite 1205, Tucson, Arizona 85750 USA
[2]Dept. of Chemical and Environmental Engineering, U. of Arizona, Tucson, Arizona 85641 USA
[3]Morgan Advanced Ceramics, 7331 William Ave, Ste. 900, Allentown, PA 18106 USA

ABSTRACT

Material removal during CMP occurs by the activation of slurry particles at contact points between pad summits and the wafer. When slurry is present and the wafer is sliding, contacts become lubricated. We present an analysis valid over the full range from static contact to hydroplaning that indicates that CMP usually operates in boundary or mixed lubrication mode at contacts and that the lubrication layer is nanometers thick. The results suggest that the sliding solid contact area is mainly responsible for the friction coefficient while both the solid contact and lubricated areas control the removal rate.

INTRODUCTION

Because of the difficulty of directly observing material removal at the asperity level in chemical-mechanical polishing (CMP), there are many unanswered questions about the fundamental mechanisms of the process. In particular, the origins of the friction force and the relationship between friction and material removal are still not well understood.

In a previous analysis [1], a compact formula from lubrication theory [2] was used to estimate the contribution μ_{visc} to the COF from Newtonian viscous shear at the contacting summits of a pad surface with an exponentially decaying summit height distribution,

$$\mu_{visc} \approx 0.9 \cdot (\mu_0 V (1 - v^2)/E)^{0.36} \kappa^{0.19} \lambda^{-0.17}. \tag{1}$$

In Eq. (1), μ_0 is the (Newtonian) slurry viscosity, V is the relative sliding speed, E is the Young's modulus of the pad, v is the Poisson ratio, κ is the mean summit curvature and λ is a measure of pad surface abruptness. While predictions from this model sometimes agree with observed power law dependences of the COF on velocity, curvature and summit height distribution, the COF from Eq. (1) is in general much smaller than observed experimentally.

Here, we examine the causes of friction and removal in CMP at a more basic level. We solve the equations of elastohydrodynamic lubrication directly and include both fluid film cavitation and mixed sliding solid contact and lubrication at contacting summits. This more detailed theory predicts that for typical ranges of pad bulk and surface characteristics, CMP operates in the boundary or mixed lubrication regime with a coefficient of friction in the observed range. In addition, the theory provides some insight into the physics underlying the COF and the removal rate.

THEORY

In elastohydrodynamic lubrication theory, the pad is treated as an isotropic elastic solid governed by the linear elasticity equations [3],

$$\partial \sigma_{ij}/\partial x_j = 0, \quad \sigma_{ij} = \lambda \varepsilon_{kk} + 2\mu \varepsilon_{ij}, \quad \varepsilon_{ij} = 1/2(\partial u_i/\partial x_j + \partial u_j/\partial x_i), \tag{2}$$

where σ_{ij} is the stress tensor, ε_{ij} is the strain tensor, u_i is the displacement in the coordinate direction x_i (alternatively x,y,z) and λ and μ are the Lame constants, $\lambda = Ev /((1+v)(1-2v))$ and $\mu = E /(2(1+v))$. In the present case, the elasticity equations are solved in a 1 mm tall column of the pad having vertical sides, a rectangular bottom, and a top surface whose topography $s=f(x,y)$ either represents an idealized summit or is extracted from confocal microscopy data. Displacements are zero on the bottom surface and are constrained to in-plane motion on the vertical sides. On the top surface, the local pressures induce pad deformation,

$$\sigma_{ij}n_j = -pn_i \tag{3}$$

where n_i are the components of the surface normal. The pressure field p includes both fluid and solid contact pressures, as described below.

The fluid pressure field p_f is found using the steady state Reynolds equation, a statement of mass conservation valid for thin films with certain additional assumptions [4,5],

$$\nabla \cdot (\rho h^3 /(12\mu_0)\nabla p_f) = \nabla \cdot (1/2\rho h \bar{V}). \tag{4}$$

Pressures are referenced to the ambient pressure, which we can take to be 0 by a change of variable. In (4), ρ is the local film density and h is the fluid thickness. The fluid thickness is related to the wafer height h_w, the surface height s and the vertical elastic displacements u_3,

$$h = h_w - s - u_3 \tag{5}$$

The elastic coupling makes (4) significantly non-linear. The wafer height is initially unknown and is determined by imposing load balance at the nominal applied pressure $\bar{p}$.

If anywhere in the domain the fluid pressure falls to the vapor pressure p_{cav}, the film cavitates. The boundary of the cavitated region, if one exists, is not known and must be determined by the solution procedure. The fluid pressure and density in the intact film must satisfy (4) as well as

$$p_f \geq p_{cav} \text{ and } \rho = \rho_{liq} \tag{6}$$

where ρ_{liq} is the normal liquid density of the slurry. In the cavitated region, (4) also applies, the pressure equals the cavitation pressure and the density must be less than the intact film density:

$$p_f = p_{cav} \text{ and } \rho < \rho_{liq}. \tag{7}$$

10

At the interface between the cavitated and intact films, the solution must also satisfy the JFO boundary conditions [6], which insure mass conservation at the cavitation boundary.

Sliding solid contact is modeled using an idea from [6], where the wafer and asperity tip are treated as rough surfaces rather than as ideally smooth ones. If the two surfaces have nanometer-scale roughness, then there will always be some fluid and fluid pressure between the pad and wafer, even in solid contact areas. The pad loses contact only if the fluid pressure exceeds the mechanical pressure from the pad. Mathematically, if the combined surface roughness creates a film of thickness h_{min} in a sliding solid contact area, then the mechanical pressures in the elasticity boundary condition (3) in the same area must be exactly those needed to produce the required minimal fluid thickness. Outside of this area, fluid pressures are used in (3). The resulting vertical pad displacements u_3 then determine the fluid thickness in (5). When the Reynolds equation (4) is solved using (5), the sliding solid contact area is treated as part of the intact film, so that at each point in this area we ultimately obtain both a solid contact pressure (from the elasticity equations) and a fluid pressure. Consistency requires that the solid contact pressure should exceed the fluid pressure.

Details of the numerical method are too intricate to fully describe here. The elasticity problem is solved by the finite element method using a 3D mesh consisting of bricks and prisms. Bricks are concentrated in the contact area. The mesh top surface nodes are probed one at a time with unit pressure in order to construct a linear operator G that relates the top surface pressures to the vertical displacements, $Gp = u_3$. This makes it possible to explicitly express the fluid thickness (5) directly in terms of the pressures. The Reynolds equation is then written entirely in terms of pressures and the coupled elasticity/flow problem solved with Newton's method. The cavitation region is determined using a method similar to the one in [5]. The surface nodes are provisionally divided between intact and cavitated regions and the appropriate form of (4) is solved in each region using the known quantities in (6) and (7) (the density in the intact region and the pressure in the cavitated region). The remaining constraints (6) and (7) are then checked and nodes that violate them are switched to the opposite region. This process is iterated until both regions pass all of the requirements. The resulting solution implicitly satisfies the JFO boundary conditions and the cavitation locus, while never explicitly gridded, clearly emerges from the algorithm. The sliding solid contact region is obtained using an analogous method.

RESULTS FOR IDEALIZED SUMMITS

We illustrate the model using idealized Gaussian summits $z = z_0 \exp(-r^2/\sigma^2)$ where $r = \sqrt{x^2 + y^2}$. These have summit curvature $\kappa = \sqrt{2z_0/\sigma^2}$. The wafer in all cases slides from left to right over a 100x100 µm area containing a single asperity and therefore an area density of $100/\text{mm}^2$. By symmetry, it is sufficient to solve on half of the domain. We use E=300 MPa, v=0.25, μ_0=2.5 mPa-s, and h_{min}=1 nm to illustrate the general behavior. Figure 1 shows typical results at and near a contact for κ=0.53/µm, z_0=17 µm, $\bar{p}$=2 PSI and V=0, 0.2, 0.3 and 0.8 m/sec. At V=0, we obtain the 2 PSI static contact area, a circular region 1.55 µm in diameter. At 0.2 m/sec, a slender lubricated crescent forms at the leading edge of the static contact area. By 0.3 m/sec, about half of the static contact area is lubricated and at 0.8 m/sec the sliding solid contact area has diminished to a small region at the lateral edge. Eventually, the sliding solid contact area goes to zero and the wafer hydroplanes. The film thickness in the lubricated area is < 3 nm and nearly constant, and the load-balanced wafer lifts by about the same amount. Even at low speeds, the film cavitates at the trailing edge of the contact. The total pressure field is constant at all speeds, fluid pressure having replaced the static solid contact pressure as the velocity increases. The shape of the sliding solid contact area and the behavior of the pressure field are well known in lubrication theory and have been verified experimentally in other contexts [7].

Figure 2(a) plots the ratio of the sliding solid contact area to the static contact area as a function of curvature and velocity for 17 µm high summits at 2 PSI. The sliding contact area shows microscopic Stribeck behavior, with boundary lubrication at low speeds, mixed lubrication at intermediate speeds and then hydroplaning. Decreasing the applied pressure moves all of the curves in Fig. 2 to the left and increasing it moves them to the right. As long as the pad is sufficiently thick, the lubrication behavior is very insensitive to the summit height z_0. The lubrication behavior is very sensitive to the minimum fluid thickness, however, with greater combined roughness decreasing the extent of lubrication (Figure 2(b)).

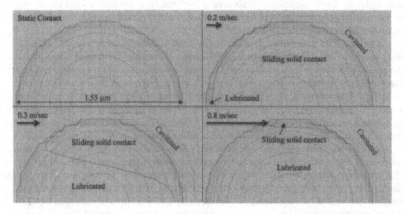

Figure 1. Lubricated and sliding solid contact regions vs. velocity for an idealized summit.

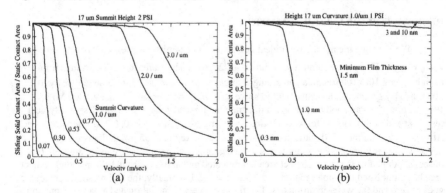

Figure 2. (a) Solid contact area vs. curvature and speed. (b) Sensitivity to surface roughness.

RESULTS FOR MEASURED SUMMITS

A 200 mm copper wafer polishing experiment was performed at 2 PSI and 1 m/sec on an Araca, Inc. APD-800 polisher and tribometer using an IC-1020 M-groove pad conditioned at 6 lb$_f$ with several types of Morgan Advanced Ceramics conditioners. Samples were removed from the pad and both surface topography and contact area at the polishing pressure were characterized using a Zeiss laser confocal microscope. Contact and topography data were taken in the same area, making it possible to see exactly which surface features are responsible for contact. Figure 3 shows corresponding contact and topography details from a pad surface produced by an advanced conditioner design. The surface is unusual in that most of the contact areas are regular and elongated. Slurry flow is from left to right in the images.

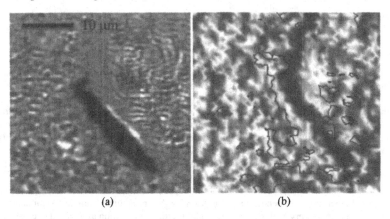

(a)   (b)

Figure 3. A static contact area (a) and the corresponding topography (b) from an IC pad. The topography is contoured at $z=11$ μm.

Topography data for the summit in Figure 3 was extracted from the larger image and smoothed to reduce measurement noise. The smoothed summit was then meshed and refined in the static contact region. Lubrication calculations were performed as above for idealized summits. Figure 4(a) illustrates the lubrication behavior of this summit. Unlike idealized summits, in which the lubricated area advances up the center of the contact, lubrication of the measured summit starts at the upstream end and advances toward the downstream end. The contact area curve also transitions much less abruptly to mixed lubrication (Figure 4(b)). Figure 4(b) also shows an estimate of the COF contribution by this contact from the solid and viscous shear forces.

DISCUSSION AND CONCLUSIONS

The above analysis, performed using parameters and conditions relevant to CMP, suggests that contacting pad summits are at least partially lubricated under most circumstances. Since the lubrication layer is much thinner than the diameter of a slurry particle, slurry particles in either the lubricated or sliding solid contact area can contribute to the removal rate. However,

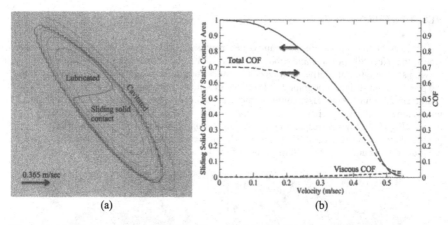

| | (a) | (b) |

Figure 4. (a) Lubrication behavior of the contact in Figure 3. (b) Calculated sliding contact area ratio and COF for the same contact as a function of velocity.

only the solid contact area contributes significantly to the COF, as seen in Figure 4(b). Removal rate and COF therefore share common but not identical physical origins. This supports the possibility that a pad could produce a high removal rate while having a very low COF. We also see from Figure 2(b) that the combined roughness of the wafer and summit can have a large effect on lubrication behavior. Nanometer-scale abrasion of the asperity tips by slurry particles or the conditioner may therefore significantly influence the COF.

Several factors have been ignored in this analysis, such as the effect of slurry particles and the possible shear rate dependence of the viscosity [8] on the lubrication layer thickness. An increase in viscosity at very high shear rates would increase lubrication and would decrease the mechanical force applied to slurry particles. Pad plasticity may also increase lubrication by decreasing curvature after several passes under the wafer and may explain the lower COF often observed on partially glazed pads. Finally, the implications of transients like wafer vibration on lubrication are as yet unexplored and may be important for explaining COF force spectra.

REFERENCES

1. L. Borucki, Y. Zhuang and A. Philipossian, Proc. 2005 PacRim Conference, pp. 411-416.
2. B.J. Hamrock and D. Dowson, Trans. ASME J. Lubrication Tech., **100**, 236-245 (1978).
3. L.D. Landau and E.M. Lifshitz, *Theory of Elasticity (3rd Ed.)*, Butterworth-Heinemann, Ch. 1.
4. A.Z. Szeri, *Fluid Film Lubrication Theory and Design*, Cambridge U. Press (1998) Ch. 2.
5. A. Kumar and J. F. Booker, Trans. ASME J. of Tribology, Vol. 113, pp. 276-286 (1991).
6. F. Shi and R.F. Salent, Trans. ASME J. of Tribology, Vol. 122, pp 308-316 (2000).
7. F. Guo and P.L. Wong, Tribology International, **37** (2004), pp. 119-127.
8. W. Lortz, F. Menzel, R. Brandes, F. Klaessig, T. Knothe and T. Shibasaki, Mat. Res. Soc. Symp. Proc. Vol. 767, F 1.7 (2003).

Multi-scale and Fundamental
Modeling of CMP

Mater. Res. Soc. Symp. Proc. Vol. 1157 © 2009 Materials Research Society   1157-E02-01

Understanding Multi Scale Pad Effects in Chemical Mechanical Planarization

Abhijit Chandra[1,2] and Ashraf. F. Bastawros[2,1]
[1]Mechanical Engineering, [2]Aerospace Engineering, Iowa State University, Ames, 50011, USA

Pavan K. Karra[1]
Mechanical Engineering, Iowa State University, Ames, IA 50011, USA

ABSTRACT
A multi-physics model encompassing chemical dissolution and mechanical abrasion effects in CMP is developed. This augments a previously developed multi-scale model accounting for both pad response and slurry behavior evolution. The augmented model is utilized to predict scratch propensity in a CMP process. The pad response delineates the interplay between the local particle level deformation and the cell level bending of the pad. The slurry agglomerates in the diffusion limited agglomeration (DLA) or reaction limited agglomeration (RLA) regime. Various nano-scale slurry properties significantly influence the spatial and temporal modulation of the material removal rate (MRR) and scratch generation characteristics. The model predictions are first validated against experimental observations. A parametric study is then undertaken. Such physically based models can be utilized to optimize slurry and pad designs to control the depth of generated scratches and their frequency of occurrence per unit area.

Keywords
Planarization, Agglomeration, Defectivity

1 INTRODUCTION
Chemical mechanical planarization (CMP) has grown rapidly during the past decade as a necessary process step in submicron integrated circuit (IC) manufacturing because of its ability to achieve global or near-global planarization [1]. Currently, CMP is widely used for interlevel dielectrics and metal layer planarization. CMP is performed by sliding the wafer surface on a relatively soft polymeric porous pad which is flooded with a chemically active slurry containing abrasive particles of sub-micron diameter. The chemical properties of the slurry interact with the mechanical properties of the abrasive particles at the nano-scale. It also affects the mechanical macro-scale properties of the polishing pad and the wafer, as well as their surface morphologies at the micro-scale. As a result, the wafer surface, the pad surface and the slurry characteristics change. This evolution controls the quality and effectiveness of the CMP process. The finished wafer surface, however, is prone to CMP process induced scratches. Existence of a single scratch whose depth is greater than a critical threshold can render the chip unusable, and this threshold continues to decrease with successive generations of IC manufacturing.
A wide range of studies on the CMP process have been reported. For example, previous work investigated the material removal rate (MRR) [2,3] and the effects of the pad and slurry properties on the process. Wang et al. [4] introduced the effects of pad wear and its evolution in an effort to extend the pad response model developed by Bastawros et al. [5], and Luo and Dornfeld [6] examined the effects of slurry properties focusing on the distribution of abrasive particles. It is well known that the slurry gradually evolves with time, with and even without continued processing, and that there is a strong correlation between slurry evolution and the

generation of scratches on the finished wafers. In the physics and colloidal chemistry communities a variety of modeling efforts as well as experimental investigations [7,8] of slurry agglomeration have been reported. There also exists a wide body of literature where the interactions between mechanical and chemical evolutions of slurry properties have been investigated [9,10]. Based on this body of work, the present study attempts to develop an integrated model of scratch generation in CMP. Evolutions of both the pad and the slurry, as well as the slurry-pad quasi-coupled interactions are accounted for and a quantitative model is developed to predict the propensity of scratches (where we quantify propensity by scratch depth and its frequency of occurrence per unit wafer area). The model predictions are compared to experimental observations and a parametric study is conducted to examine the effects of the pad and wafer surface hardness on scratch propensity.

2 AGGOLMERATION AND SLURRY EVOLUTION

When particles come into contact due to Brownian motion they tend to stick to one another to form agglomerates. Although the agglomeration process is traditionally classified into two separate regimes, diffusion limited agglomeration (DLA) and reaction limited agglomeration (RLA), these two regimes can be considered as variations due to slurry pH and other chemical factors within a single generalized regime. DLA occurs near the iso-electric point of the slurry particles. The iso-electric point refers to the pH of the slurry at which the particles have zero electro-kinetic potential, which implies that the charge on their surface is zero [10]. Since there is no barrier to keep the particles apart they tend to stick under a single collision.

By contrast, RLA occurs when the pH of the slurry is away from the iso-electric point. Here a net charge on the surface of the particles exists. This charge, which depends on pH, creates an electro-kinetic potential between the particles which they must overcome to stick to each other. Their ability to penetrate this barrier depends on their kinetic energy $k_B T$, pH and other electrochemical parameters, as well as the strength of the van der Waals forces. Since van der Waals forces act over a shorter range when compared to the separation between particles, the primary cause for overcoming the barrier is kinetic energy. Keeping other parameters constant, the higher the kinetic energy, the higher will be the probability of agglomeration. The van der Waals forces however become effective when the particles contact each other, and prevent them from separating. In the present study, we investigate the case when the slurry pH is close to its iso-electric point, i.e., the DLA regime.

For the general case, the agglomeration process can be modeled using the Smoluchowski rate equation [9] which gives the time rate of change of the number of particle clusters with volume M, $N(M)$ as

$$\frac{d}{dt}N(M) = \frac{1}{2}\sum_{K=1}^{M-1} a(M,K)N(M-K)N(K) - \sum_{K=1}^{\infty} a(M,K)N(M)N(K) \tag{1}$$

The agglomeration kernel, $a(M,K)$ is the rate at which clusters of volume M agglomerate with particles of volume K. It has been shown that most agglomeration results from smaller particles adding themselves onto larger clusters [9], i.e., $M>>K$. For the DLA regime, and for $M>>K$, $a(M,K)$ reduces to

18

$$a = \frac{8k_B T N_0}{3\eta} \tag{2}$$

where k_B is the Boltzmann constant, T is absolute temperature, N_0 is the initial number of particles and η is the viscosity of the slurry. Substituting Equation (2) into Equation (1), the asymptotic solution for the final volume distribution of clusters can be found. The time to reach the asymptotic volume distribution depends on Equation (2), but this volume distribution is found to be independent of the initial conditions as long as the particles are sufficiently small [7,8]. The particle cluster volume distribution can then be used to obtain an equivalent particle cluster radius distribution, X. Particles are known to agglomerate with fractal dimensions [7]. This implies that cluster volume is not proportional to the cube of the cluster radius, (i.e., $V = 4/3\pi X^3$) but to the power of the fractal dimension. Based on detailed experiments, Lin et al. [7] have observed the fractal dimension to be from 1.86 to about 2 for DLA. Accordingly, in our model calculations, we utilize this range to convert the volume of a particle cluster into its equivalent radius.

3 PAD EVOLUTION

Pad evolution plays an important role in the CMP process. It has been observed that MRR drops as the pad wears out. The spatial distribution of the MRR is also affected by pad wear, which takes place mainly at the asperity level. Pad asperity evolution has been quantified by Wang et al. [4] for the case where the asperity height z is greater than the average separation between the pad and the wafer $d(t)$. The probability density function (PDF) of the asperity height z at any time t is given as

$$\frac{d}{dt}\phi(z,t) = \frac{4C_a E^* \sqrt{\kappa_e}}{3\pi} \frac{\partial}{\partial z}\left\{\sqrt{z - d(t)}\phi(z,t)\right\} \tag{3}$$

where C_a is the pad wear rate coefficient, E^* is effective modulus of the pad (for this case, approximately equal to pad modulus) and κ_e is equivalent curvature. The calculation of equivalent curvature via a multi-physics aggregation algorithm is explained in section 5. Equation (3) considers the evolution of the asperities whose height is greater than $d(t)$, and therefore contact the wafer. The separation between wafer and pad $d(t)$ is also a function of time and changes as the asperities wear. Equation (3) is subject to the initial condition $(\phi(z,t=0) = \phi(z,0) = \phi_0(z)$. Based on the surface profile of an IC 1000 pad [4] a Pearson-IV PDF is used in the present study as the initial pad asperity height PDF. The scratch depth calculation described in the next section is significantly influenced by the instantaneous distribution of asperities on the pad surface.

4 SCRATCH DEPTH

The principal material removal mechanism is plowing of particles on the softened wafer surface. Plowing involves indentation followed by dragging of the slurry particle. Indentation takes places as particles are entrapped between the pad asperities and the wafer surface. The particle size is typically about 100 nm, while asperity height and tip radius are usually several micrometers. Therefore hundreds of particles are trapped between each asperity and the wafer surface. The calculation of scratch depth involves two random variables, pad asperity height z and effective particle cluster radius X (as calculated using the cluster volume distribution and the fractal dimension of the cluster). The two variables are independent and the scratch depth $W(i,j)$ due to j^{th} particle under i^{th} asperity is given by [11] as

$$W(i,j) = \frac{E^* \sqrt{\kappa_e}}{H\pi} \sqrt{z - d(t)} X(i,j)$$ (4)

Using Equation (4) the cumulative density function (CDF) of the scratch depth is calculated. The calculation of equivalent curvature κ_e via a multi-physics aggregation algorithm is explained in section 5. $P(W \leq w)$ is the probability per active particle that a scratch of depth W, which is less than a prescribed threshold w, will be created and is given as

$$P(W \leq w) = \int_0^{X_{max}} \int_0^{\frac{w^2}{H^2}} f_z(z) f_x(x) dz' dx$$ (5)

Here, $f_{Z'}$ is the PDF of the asperity height and f_X is the PDF of the particle radius distribution. In the above $z' = z_i - d(t)$ and H is expressed in terms of the hardness of the wafer H_w, the effective modulus of the pad E^* and the equivalent curvature κ_e as

$$H = \frac{3\pi H_w}{2E^* \sqrt{\kappa_e}}$$ (6)

In the present study, a hardness of 4 GPa was used for the softened (hydroxylated) layer of the wafer (Si(OH)) [12].

5 MULTI-PHYSICS INTERACTION OF CHEMICAL AND MECHANICAL PHENOMENON

In the indentation model, the pad asperity curvature plays a significant role. This is a mechanical aspect that affects the surface evolution in CMP. CMP also involves a chemical component via the chemical dissolution effects and according to [13] there is a critical wavelength of λ_{cr} (0.7 micron for the present case) that is imposed by the chemical dissolution of the wafer due to the slurry. The amplitude at λ_{cr} is observed to be 20 micron. This profile can be used to obtain a surface curvature induced by the chemical dissolution effects.

A multi-physics approach can now be employed to address the combination of these two (chemical and mechanical) effects on scratch propensity in CMP. The algorithm used in this work follows from [14] and [15]. This algorithm is used here to calculate the equivalent curvature that accounts for both the chemical and mechanical aspects described above.

The curvature matrix is composed of chemical and mechanical aspects. Both mechanical and chemical effects have short term and long term effects as expressed below.

$$\kappa_c = \kappa_c^S \kappa_c^L; \quad \kappa_m = \kappa_m^S \kappa_m^L$$

20

The subscripts c and m stand for chemical and mechanical, respectively, while the superscripts S and L stand for short term and long term, respectively.

The equivalent curvature of chemical and mechanical effects, which contain long term and short term effects, can be described as [15]

$$\kappa_e = \begin{bmatrix} \kappa_c^S \kappa_c^L & \alpha \kappa_c^S \kappa_m^L \\ \alpha \kappa_m^S \kappa_c^L & \kappa_m^s \kappa_m^L \end{bmatrix} \tag{7}$$

Recognizng the fact that both chemical and mechanical curvatures are imposed at the same location or geometrical point, the scalar effective (or equivalent) curvature can be obtained by the sum of individual elements of the matrix in equation (7). If the chemical reaction rate is slow relative to mechanical response rate, the chemical effects are predominantly long term which implies that the short term chemical effects are negligible, i.e. $\kappa_c^s = 0$. For the mechanical effect the short term and long term effects are approximately same. Thus,

$$\kappa_m^S = \kappa_m^L \tag{8}$$

The scalar effective curvature now becomes

$$\sqrt{\kappa_m^s \kappa_m^L + \alpha \kappa_m^s \kappa_c^L} \tag{9}$$

The parameter κ_e in the equation (7), and (9) determines the extent to which chemical and mechanical effects interact with each other in CMP. The parameter α is the coupling coefficient, which can be determined by fitting a single experimental data point. The limits of α are between -1 and 1, implying fully collaborative or fully competitive interactions.

6 EXPERIMENTAL PROCEDURE

The CMP experiments were carried out at IMEC in Belgium and the experimental parameters measures are listed in table 1. Most of the parameters are measured for this experiments and the rest are taken from literature. For all experiments, 200 mm silicon wafers with a 500 nm high-density plasma (HDP) oxide film deposited on the surface were used. The polishing tool was a rotary polisher (E460 Mecapol polisher, Alpsitec). The experiments were performed using an IC 1000 pad. The pressure was kept at 28 kPa, and the slurry volume particle concentration was 5% colloidal silica with an initial mean diameter of 100 nm and a standard deviation of 30 nm. The slurry was maintained at a pH of 6. The slurry was prepared and the pH adjusted about 2 days before using it in the CMP experiments. This allowed the slurry particles enough time to achieve conditions consistent with asymptotic DLA agglomeration. The pad and wafer rotational speeds were 65 RPM and 40 RPM respectively, and the pad diameter was 0.5 m. Polishing was performed for 1 min at 300 K using a slurry flow rate of 200 ml/min. The scratch depth and frequency of occurrence on the polished wafer was measured by atomic force microscopy over a 10μm x 10μm area. For a given scratch, the depth was measured at three locations along the scratch and the largest depth reported [16]. Similar experiments were done by Gopal and Talbot [17].

	Value	Units
Pad Modulus	29	MPa
Asperity Curvature	.02	/micron
Particle mean Dia	100	nm
Particle Std Deviation Dia	30	nm
Pad Speed	65	RPM
Platen Speed	40	RPM
Slurry particle concentration	5%	vol
Slurry flow rate	200	ml/min
pH of slurry	6	
Polishing time	1	min
Pad Diameter	0.5	m
Slurry Viscosity-Water	0.001	Pa.s
Temperature	300	K
Hydrated Wafer Hardness	4	GPa
Fractal Dimension	1.86 - 2.1	

Table 1: List of parameters used in the simulation. The table lists the sources of the parameters

7 RESULTS AND DISCUSSION

The model was used to predict the scratch depth and frequency of occurrence per 100 μm x 100 μm wafer area for the same conditions present in the experiments. In addition, a pad modulus of 29 MPa [4], an asperity curvature of 0.02 μm^{-1} [4] and a slurry viscosity of 0.001 Pa·s were used. Equations (1), (3) and (4) were solved simultaneously to obtain the scratch depth. To obtain the frequency of occurrence, $P(W \leq w)$ in Equation (5) is multiplied by the number of active particles used. The fraction of pad area in contact with the wafer determines the number of active particles. A value of $A_f = 0.008$ [4] was used, which is consistent with the pressure used in the experiments [16]. The simulation was run for 3000 s at which asymptotic DLA agglomeration conditions had been reached.

The calculation of curvature is explained in section 5. Based on these calculations simulations were run for entire mechanical, entire chemical and coupled chemical-mechanical conditions. The scratch propensity is then calculated for two different fractal dimensions of 1.86 and 2.1. The model predictions are compared with the experimental results in Figure 1. No other fitting parameters are used for these comparisons. Here, rather than fitting α with a single experimental observation, the limits of model predictions are explored with α varying between -1 and 1. The histogram represents the experimental observations. The red and blue curves are pure mechanical or pure chemical effects with fractal dimensions of 1.86 and 2. The purple line represents $\alpha = -1$, and the green line represents $\alpha = 1$ at different fractal dimensions of 1.86 and 2.

It is interesting to note that the comparison to experimental observations improve significantly as α is varied from +1 to -1. This is expected, since in CMP, the internal interactions of curvatures remain on the evolved surface, but the external interactions occur at asperity peaks and are quickly wiped out as polishing debri. It is also significant to note that the experimentally observed most frequent peak is at around 250 nm, which is still lower than the 300 nm position obtained with $\alpha = -1$. This happens because the change in curvature due to the mechanical compliance of the pad under a given load is not accounted for in the current model. The results indicate that the pad used in the experimental study is a compliant pad with significant deflection at asperity tips. Such behavior may be an indication of aggressive conditioning strategy that may have been used.

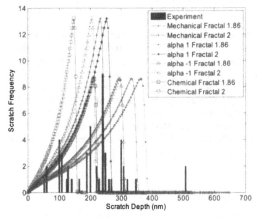

Figure 1: Comparisons of model predictions for scratch propensity with experimental observations; The "mechanical" implies that only mechanical effects are considered and the label "chemical" implies that only chemical effects are considered. The label "alpha" is the fit for the value of alpha mentioned.

The maximum scratch depth found in the experiments was about 500 nm with a frequency of occurrence of 2. The maximum scratch depth predicted by the model is much lower. This discrepancy is thought to be caused by inaccuracies in the assumed initial particle distribution in the slurry. This could occur if the slurry had been contaminated by a very small percentage of relatively large particles (compared to the mean particle size of 100 nm). Otherwise, it is observed that the scratch depths and scratch frequencies predicted by the model are quite comparable to experimental observations. In the experiments, a maximum scratch frequency of 9 is observed for a scratch depth of around 250 nm. For a fractal dimension of 2, the model predicts a maximum scratch depth of 250 nm with a frequency of about 13. At the other limit, taking the fractal dimension of 1.86, a scratch frequency of 9 is predicted for a scratch depth of around 350 nm. It is seen that the model correctly predicts either the scratch frequency of 9 (for a fractal dimension of 1.86), or the correct scratch depth of 250 nm (for a fractal dimension of 2). However, it fails to make simultaneous correct predictions for both the quantities defining the scratch propensity. We believe this is caused by using an approximate agglomeration fractal dimension. To obtain an accurate estimate of the full scratch propensity (both scratch depth and frequency), a detailed experiment may be needed to obtain the exact fractal dimension for the experimental conditions used. Such measurements can only be made a posteriori however, and therefore cannot be part of any predictive strategy.

8 PARAMETRIC STUDY

The model is verified against the experiment and the model prediction agrees well with the experiment. There are different variables that have effect on the scratch depth and these variables play an important role in the final surface quality of the wafer surface because these parameters are one of the main variables that define the process. In this paper effect of pad modulus and the effect of wafer surface hardness are studied.

8.1 Pad effect

In a typical CMP process various properties of pad play an important role in determining material removal process. One main property is pad modulus. Pad modulus changes with the process conditions such as slurry feed rate and temperature of the process. Slurry feed rate determines the wetness of the pad which in turn determines the pad modulus. Pad modulus ranges from 29 MPa for a wet pad till 100 MPa for a dry pad. It can be seen from equation (4) that pad modulus affects the indentation depth hence the probability density of indentation depth. This trend can be seen in figure 2. As the pad modulus increases the indentation depth distribution moves to the right and with it the mean indentation depth also increases. This increases the material removal rate and also the defectivity since there are more number of deep scratches for a dry pad.

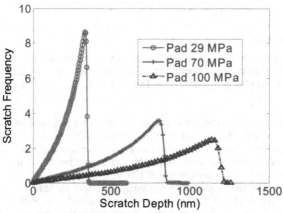

Figure 2: Effect of Pad Modulus

8.2 Wafer surface hardness

Different CMP application polish different material present on the surface which means that the wafer surface hardness changes with process. The process parameters such as pressure, slurry concentration slurry pH and velocity need to be adjusted with respect to the material to be removed from the wafer surface. In oxide CMP the wafer surface is wetted by slurry and the hydrated wafer surface is softer (hardness 4 GPa) than dry wafer surface (7 GPa) which occurs due to insufficient slurry solution. In copper CMP copper layer is removed from the wafer surface and copper has a hardness of 800 MPa which is much softer than the oxide. This means that indentation distribution moves to right compared to oxide CMP and the mean removal rate is higher for same process conditions. This is reflected in Figure 3 as predicted by equation (6).

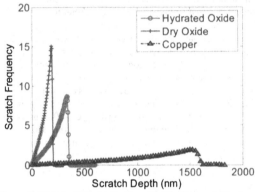

Figure 3: Effect of Wafer Surface (Oxide/Copper) Hardness

It has to be noted that the model prediction of the size distribution is slightly higher than the experimental result which is evident from the figure as the depths of maximum probability for model and experiment differ by around 50 nm. In a typical CMP process the slurry particles are also softened along with the wafer surface. This is particularly true in the present case since the wafer is made of same material as the slurry. In the indentation model described earlier the slurry particle is considered to be rigid which causes the model to predict higher than experimental result. Relaxing the assumption of rigid particles and taking particle stiffness into consideration will alleviate the evident disparity.

8 CONCLUSIONS

Scratching is an essential part of the material removal mechanism in CMP. MRR, and its associated productivity, is significantly reduced if the scratch generation process is hindered in any way. However, it is the scratch generation characteristics that determine the quality of the finished surface. In IC manufacturing, an entire chip may be rendered unusable if a "killer" defect (often defined as a scratch deeper than the specified limit) is present. Thus, a propensity for shallower but more numerous scratches may be preferred over the detrimental tendency to produce fewer but deeper scratches.

The present study provides an integrated modeling strategy that combines both slurry evolution and pad evolution to predict scratch generation propensity for a CMP process. It is observed that the fractal dimension of the slurry agglomeration process plays a central role in determining the scratch generation propensity, and the inability to determine this dimension a priori significantly limits the predictive success of any modeling effort. In the present analysis, we have utilized two previously established limits of fractal dimensions. This allows us to predict either maximum scratch depth or its associated frequency accurately. However, we were unable to make accurate and simultaneous determination of both of these quantities. Thus, the value of the present modeling effort lies in establishing limiting cases. More accurate comparisons to experimental observations may be made either by fitting α to a single experimental data point, or by accounting for the fact that the mechanicl curvature at asperity tips are significantly modified under loding due to its inherent compliance. Finally, it should be noted that neither the mechanical curvature, nor the chemical curvature is a single value (as assumed in this study for simplicity), but, in reality, represents a distribution that need to be accounted for. We believe that accounting for such curvature distributions is necessary, and will significantly improve the comparison of model prediction to the full experimental distribution of scratch propensity.

It is observed that scratch depth increases while scratch frequency decreases for harder (drier) pads as well as for softer surface layers on wafer surfaces. Thus, an over aggressive slurry chemistry, which significantly reduces the surface layer's hardness, may be equivalent to using harder pads in a CMP process. Similarly, the need for scratch "free" finishing of softer materials (e.g., copper) may be addressed by making sure that the agglomeration process in the slurry is inhibited (by staying away from the iso-electric point of the slurry), by using softer pads and by designing slurry flow patterns that keep the pad fully and continuously wet.

9 ACKNOWLEDGEMENTS

The authors gratefully acknowledge the support of NSF under Grant CMMI-0640826. Any opinions, conclusions or recommendations expressed are those of the authors and do not necessarily reflect views of the sponsoring agencies. The authors also wish to thank Prof. D. G. Saari for many helpful discussions on multi-scale and multi-physics analysis.

10 REFERENCES

[1] Martinez, M.A, 1994, Solid State Technology.

[2] Komanduri R, 1996, On material removal mechanisms in finishing of advanced ceramics and glasses, Annals of the CIRP, 45:509-514.

[3] Evans, C.J., Paul, E., Dornfeld, D., Lucca, D.A., Byrne, G., Tricard, M., Klocke, F., Dambon, O., Mullany, B.A, 2003, Material removal mechanisms in lapping and polishing, Annals of the CIRP, 52/2:611-634.

[4] Wang, C., Sherman, P., Chandra, A., Dornfeld, D., 2005, Pad Surface Roughness and Slurry Particle Size Distribution Effects on Material Removal Rate in Chemical Mechanical Planarization, Annals of the CIRP, 54/1:309-312.

[5] Bastawros, A.F., Chandra, A., Guo, Y., Yan, B., 2002, Pad Effects on Material Removal Rate in Chemical Mechanical Planarization, J. Electronic Materials, 31/10:1-10.

[6] Luo, J., Dornfeld, D., 2003, Effects of abrasive size distribution in chemical mechanical planarization: modeling and verification, IEEE Trans. Semiconductor Manufacturing, 16/3:469-476.

[7] Lin, M Y., Lindsay, H.M., Weitz, D.A., Klein, R., Ball R.C., Meakin, P., 1989, Universality of Fractal Aggregates as Probed by Light Scattering, Proceedings of the Royal Society of London A, 423/1864:71-87.

[8] Lin, M Y., Lindsay, H.M., Weitz, D.A., Klein, R., Ball R.C., Meakin, P., 1990, Universal diffusion-limited colloid aggregation, Journal of Physics: Condensed Matter, 2:3093-3113.

[9] Ball, R.C., Weitz, D.A., Witten, T.A., Leyvraz, F., 1987, Universal kinetics in reaction limited aggregation, Physical Review Letters, 58/3.

10] Komulski, Marek, 2001, Chemical properties of material surfaces, New York.

[11] Che, W., Guo, Y.J., Chandra, A., Bastawros, A., 2005, A Scratch Intersection Model of Material Removal During Chemical Mechanical Planarization (CMP), Journal of Manufacturing Science and Engineering, 127/3:545-554.

[12] Bastaninejad, M., Goodarz, A., 2005, Modeling the Effects of Abrasive Size Distribution, Adhesion and Surface Plastic Deformation on Chemical-Mechanical Polishing, J. of The Electrochemical Society, 152/9:1875-1877.

[13] Che W, Bastawros A, Chandra A, et al., Surface evolution during the chemical mechanical planarization of copper, CIRP Annals-Manufacturing Technology, Volume: 55, Issue: 1, Pages: 605-608 Published: 2006.

[14] Chandra, A., P, Karra and M, Dorothy., Implications of Arrow's Theorem in Modeling of Multiscale Phenomena: An Engineering Approach, Submitted to Journal of Design, also presented at NSF CMMI Grantee conference, Knoxville, Jan 2008.

[15] Chandra, A., Unpublished work, 2009.

[16] Armini, S., 2007, Composite particles for chemical-mechanical planarization applications, Dissertation, IMEC, Kapeldreef 75, Leuven, Belgium.

[17] Gopal, T., Talbot, J.B., Use of Slurry Colloidal Behavior in Modeling of Material Removal Rates in CMP, Journal of The Electrochemical Society, 154/6, H507-H511, 2007.

[18] Boroucki, L,. Mathematical Modeling of Polish-rate Decay in Chemical-Mechanical Polishing, Journal of Engineering Mathematics 43, 105–114, 2002.

Mater. Res. Soc. Symp. Proc. Vol. 1157 © 2009 Materials Research Society   1157-E02-02

A Study to Estimate the Number of Active Particles in CMP

Jeremiah N. Mpagazehe[1], Geo Thukalil[1], C. Fred Higgs III[1]
[1]Carnegie Mellon University, 5000 Forbes Avenue, Pittsburgh, PA 15213

ABSTRACT

To improve yield rates during integrated circuit fabrication, a better understanding of the material removal process during CMP is sought. Many material removal models have been generated to predict the material removal rate (MRR) during CMP. The majority of such models estimate that the MRR is equal to the material removed by a single particle multiplied by the total number of particles contributing to the wear process. Particles contributing to the wear process are known as 'active particles'. While several authors have proposed analytical models to estimate this quantity, this work introduces a new method for estimating the number of active particles in CMP by deducing it from the polish results of a multi-physics CMP model. By employing The Particle-Augmented Mixed Lubrication model (PAML) developed by Terrell and Higgs (2008), it is possible to determine the number of active particles in CMP. The predictions of PAML are compared with two popular analytical approaches which have been commonly used to predict the number of active particles during CMP.

INTRODUCTION

CMP is the primary method for planarization in integrated circuit fabrication. A greater understanding of CMP is desired as inconsistent polishing leads to degradation in circuit performance. Many authors have posited models to predict material removal rates (MRRs) during CMP [1],[6],[9]. However, these material removal models have employed a wide range of physical phenomena to predict wear mechanisms. Strikingly, despite the range of hypotheses and approximations developed to predict wear, many of the methods used to model CMP base the MRR formulation on the number of particles which take part in the wear process, multiplied by the volume per time removed by each of these particles. The number of particles taking part in the wear process is defined as N_a.

$$MRR_{global} = N_a \times \frac{dVol_{local}}{dt} \qquad \text{Eqn. (1)}$$

The purpose of this work is to gain a better understanding of N_a by comparing the results from several models. Two analytical models were selected because of their difference in underlying assumptions and their pervasive use in the field of MRR prediction. A third model is a new computational approach which can, for the first time, provide an *in-situ* prediction of the number of active particles during a multi-physics CMP simulation. Each of these models predicts the number of active particles in a fundamentally different manner. The resultant number of active particles is a difficult parameter to verify experimentally. Thus, in the absence of experimental results which directly measure N_a, the comparison of these three different approaches to predict N_a will help to verify this common parameter in CMP modeling.

Approaches to Predict N_a

This work compares three different techniques for predicting the number of active particles during CMP: The Particle Distribution Technique, The Particle Mono-Layer Technique, and The Particle-Augmented Mixed Lubrication (PAML) model. Fig. 1 displays characteristic differences in the particle's interaction with the wafer and the pad for each approach.

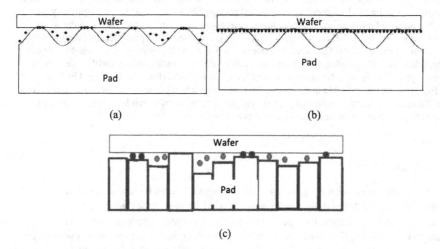

(a)   (b)

(c)

Figure 1: Characteristics of each N_a model
(a) *The Particle Distribution Technique*
(b) *The Particle Mono-Layer Technique*
(c) *The Particle-Augmented Mixed Lubrication (PAML) model.*

The Particle Distribution Technique:

Introduced by Luo and Dornfeld in 2001 [6], The Particle Distribution Technique assumes particles of varying diameter remove material from the wafer. In this technique, the number of particles sandwiched between the real contact area of the wafer and the pad is defined as N^*. The pad's asperities are assumed to have a uniform periodic distribution. As the particles are pressed into the wafer, the pad's asperities deflect around the largest particles. Only the particles which have a diameter larger than the separation distance between the wafer and the deflected pad sustain a load. A probability distribution function, based on the distribution of particle sizes, material properties and CMP parameters is used to determine the percentage of N^* which will sustain load. These particles are considered active as only the particles which are pressed into the wafer's surface contribute to the wear.

$$N_a = N^* \left[1 - \left(\begin{array}{c} \textit{Probability that} \\ \textit{a particle is smaller than} \\ \textit{the seperation} \\ \textit{between the wafer} \\ \textit{and pad} \end{array} \right) \right] \qquad \text{Eqn. (2)}$$

N^* is calculated based on the volume of the asperities contacting the wafer.

The Particle Mono-Layer Technique:

Originally developed by Zhao and Chang in 2002 [2], The Particle Mono-Layer Technique is explored as an alternative method to predict N_a. In this method, particles are assumed to be distributed uniformly across the wafer-pad interface. The active particles are characterized by the particles which happen to be sandwiched between the pad and the wafer as a result of real contact. The remaining, inactive particles, reside in difference between the apparent contact area and the real contact area. Zhao and Chang's formulation for the number of active particles is reproduced in Eqn. 3.

$$N_a = A_{real} \left(\frac{6 \, SF}{\pi D^3} \right)^{2/3} \qquad \text{Eqn. (3)}$$

A_{real} represents the real contact area, SF represents the solid fraction (volume-basis) and D represents the average particle diameter. The term multiplied by A_{real} is the areal density of the particle mono-layer.

The Particle-Augmented Mixed Lubrication (PAML) Model:

As an alternative to the analytical Particle Distribution and Particle Mono-Layer Techniques, a computational approach is explored. In 2008 Terrell and Higgs introduced The Particle-Augmented Mixed Lubrication (PAML) model [4]. PAML is a multi-physics computational model which combines solid-mechanics, fluid-mechanics, and particle-dynamics to predict wear during CMP. Solid-Mechanics in PAML is simulated using a Winkler Elastic-Foundation applied to pixilated volume elements called voxels. Fluid-Mechanics in PAML is simulated using an in-house three-dimensional CFD solver. Momentum exchanged in particle-particle and particle-surface collisions is calculated using an in-house particle-dynamics solver. Much like The Particle Distribution and Particle Mono-Layer Techniques, PAML designates active particles as particles which are loaded between the asperities of the pad and the asperities of the wafer. However, in contrast to the analytical techniques, the locations of these particles are not assumed but are determined as a result of first-principle physics employed by the solid-mechanics, fluid-mechanics, and particle-dynamics modules. PAML simulates the motion of the fluid particles due to viscous drag with the slurry fluid, and also the reaction forces on the fluid as the particles are accelerated. These aspects of PAML make it fundamentally different from the previous two analytical techniques in determining N_a.

DISCUSSION

Each approach yields a prediction for the number of active particles during CMP. The relevant parameters for the comparison are summarized in Table 1.

Table 1: Relevant Parameters for N_a Comparison

Relevant Parameters	Value
Particle Diameter (nm)	286
Slurry Solid Fraction	0.04
Nominal Contact Area (μm^2)	625
Real Contact Area (μm^2)	4.88
*Real Contact Area Particle Distribution Technique (μm^2)	136.72

For the Particle Mono-Layer Technique and PAML the pad surface was generated from a prescribed Gaussian asperity distribution with a mean asperity height of 10 μm and a standard deviation of asperity heights of 5 μm. To make the comparison with The Particle Distribution Technique as presented in Luo and Dornfeld 2001 [6] special care had to be taken with regard to the pad asperity distribution. Thus, in this study of The Particle Distribution Technique, the pad surface is assumed to be a uniform periodic distribution of the same mean asperity height and standard deviation as prescribed in The Particle Mono-Layer technique and PAML. As a result, the real contact area for the Particle Distribution Approach is signifcantly larger as depicted in Table 1. The standard deviation of particle sizes for all three models is zero. With no deviation in particle size, The Distributed Particle Technique will result in a N_a prediction of N*. The N* values that resulted from the modified pad topography are multiplied by a scaling percentage of 0.2% which is found to be typical of the application of The Distributed Particle Technique for real slurries in CMP [8]. The N_a Predictions from all three models are presented in Table 2.

Table 2: N_a Predictions

Approach Name	N_a Predicted
Particle Distribution Technique	7.55
Particle Mono-Layer Technique	10.74
PAML	10.49

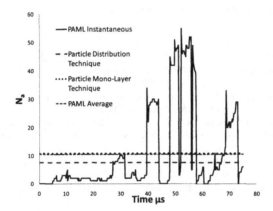

Figure 1: Comparison of N_a predictions

Because PAML is a transient simulation the number of active particles varies with time. The instantaneous number of active particles can be seen in Figure 1. For a comparison with the other models, PAML's instantaneous N_a prediction is averaged over the simulation's duration. The predicted number of active particles during CMP for all three models is very close. The N_a predictions of The Particle Mono-Layer Technique and PAML are nearly identical. The PAML modeling results could be improved by employing longer simulation times, more uniform pad topographies, and commercially relevant statistical particle distributions.

CONCLUSION

This work compared the results of N_a predictions from three models. The Distributed Particle Technique and The Particle Mono-Layer are analytical approaches while PAML is a multi-physics computational CMP simulation. It was found that the models give very similar predictions for the number of active particles during CMP. However, for The Particle Distribution technique, the underlying assumptions of the pad topography and the particle size distribution had to be accounted for while such adjustments were not made for the comparison of

The Particle Mono-Layer Technique and PAML. These results suggest that a multi-physics CMP model such as PAML may be a powerful tool for predicting key parameters such as the number of particles active during the CMP process. While the accuracy of the prediction can be enhanced by using more complex pad and particle parameters, this work shows that rigorous multi-physics computational modeling approaches of the CMP process may be a suitable for validating simpler closed-form MRR relations.

ACKNOWLEDGMENTS

The authors gratefully acknowledge the National Science Foundation CAREER program in the CMMI directorate for their support of this work.

REFERENCES

[1] Goodarz Ahmadi, Xun Xia, "A model for Mechanical Wear and Abrasive Particle Adhesion during the Chemical Mechanical Polishing Process," *Journal of The Electrochemical Society*, vol. 148, pp. 99-109, 2001.

[2] Yongwu Zhao, L. Chang, "A micro-contact and wear model for chemical-mechanical polishing of silicon wafers," *WEAR*, vol.252, 2001.

[3] D. Castillo-Mejia, S. Beaudoin, "A Locally Relevant Wafer-Scale Model for CMP of Silicon Dioxide," *Journal of The Electrochemical Society*, vol. 150, pp. 581-586, 2003.

[4] E. J. Terrell and C. F. Higgs, "A Particle-Augmented Mixed Lubrication Modeling Approach to Predicting Chemical Mechanical Polishing," *Journal of Tribology-Transactions of the Asme*, vol. 131, p. 10, Jan 2009.

[5] S. Chandrasekar, *et al.*, " Mechanics of Polishing," Transactions of the ASME, vol.65, 1998.

[6] J. F. Luo and D. A. Dornfeld, "Material removal mechanism in chemical mechanical polishing: Theory and modeling," *Ieee Transactions on Semiconductor Manufacturing*, vol. 14, pp. 112-133, May 2001

[7] Ashraf Bastawros, Abhijit Chandra, Yongjin Gao, Bo Yan, "Pad Effects on Material-Removal Rat in Chemical-Mechanical Planarization", *Journal of Electronic Materials*, vol. 31, No. 10, 2002.

[8] J.F. Luo and D.A. Dornfeld, "Effects of Abrasive Size Distribution in Chemical Mechanical Planarization: Modeling and Verification" *Ieee Transactions on Semiconductor Manufacturing*, vol. 16, pp. 469-476, August 2003

[9] W. Che, *et al.*, "Mechanistic understanding of material detachment during micro-scale polishing," *Journal of Manufacturing Science and Engineering-Transactions of the Asme*, vol. 125, pp. 731-735, 2003.

Mater. Res. Soc. Symp. Proc. Vol. 1157 © 2009 Materials Research Society   1157-E02-03

Integrated Tribo-Chemical Modeling of Copper CMP

Shantanu Tripathi[1], Seungchoun Choi[1], Fiona M. Doyle[2], and David A. Dornfeld[1]
[1]Department of Mechanical Engineering, University of California,
Berkeley, CA 94720-1740, U.S.A.
[2]Department of Materials Science and Engineering, University of California,
Berkeley, CA 94720-1760, U.S.A.

ABSTRACT

Copper CMP is a corrosion-wear process, in which mechanical and chemical-electrochemical phenomena interact synergistically. Existing models generally treat copper CMP as a corrosion enhanced wear process. However, the underlying mechanisms suggest that copper CMP would be better modeled as a wear enhanced corrosion process, where intermittent asperity/abrasive action enhances the local oxidation rate, and is followed by time-dependent passivation of copper. In this work an integrated tribo-chemical model of material removal at the asperity/abrasive scale was developed. Abrasive and pad properties, process parameters, and slurry chemistry are all considered. Three important components of this model are the passivation kinetics of copper in CMP slurry chemicals; the mechanical response of protective films on copper; and the interaction frequency of copper with abrasives/pad asperities. The material removal rate during copper CMP was simulated using the tribo-chemical model, using input parameters obtained experimentally in accompanying research or from the literature.

INTRODUCTION

Any mechanistic model of copper CMP must capture the synergistic interaction of mechanical and chemical/electrochemical phenomena. The overall corrosive wear rate, V_{CW}, exceeds the sum of the material loss rates observed in pure corrosion, V_C, or pure wear in the absence of corrosion, V_W[1]. This excess removal rate is due to the combined effect of corrosion-induced wear, ΔV_W, and wear-induced corrosion, ΔV_C. The former results from phenomena such as abrasion by hard, oxidized wear products; the removal of work hardened layers by oxidative dissolution, or stress-corrosion cracking. The latter, wear-induced corrosion, can be due to enhanced corrosion after mechanical removal of protective films, the enhanced activity due to strained surfaces or other defects, local increases in temperature or improved mass transport.

Direct wear of metallic copper would not be expected during CMP, as has been observed experimentally[2,3], hence V_W is zero. Slurries are formulated to minimize direct dissolution of copper, V_C, in the absence of mechanical action, to prevent attack of recessed regions of the wafer surface. Planarization can only be achieved if $V_{CW} >> V_C$.

Much of the existing copper CMP research considers the difference between the total removal rate during CMP and the rate of dissolution in the absence of polishing to be mechanical (corrosion enhanced wear, ΔV_W), attributed to modification of hardness and/or Young's modulus by the chemical environment. This generates material removal models that are primarily mechanical, with empirical inputs from the chemical processes. This approach provides no route for fully describing the mechanisms thought to be involved in CMP.

From a mechanistic perspective, passive or protective films containing oxidized copper are mechanically removed (or partially removed) during CMP by pad-abrasive action. This leaves a more reactive surface on which the protective layer starts to reform, until it is removed by the next abrasive particle or pad asperity. This regular removal of the protective film leads to overall oxidation rates significantly higher than those that would occur without mechanical disruption. Hence this phenomenon is best described as wear-enhanced corrosion, ΔV_C, rather than the corrosion enhanced wear, ΔV_W currently considered in most CMP models. Since only oxidized copper is removed during CMP (dissolved or abrasion of films), the most appropriate way to model this is electrochemically, using oxidizing currents along with inputs that describe the mechanical phenomena. Here we present a new modeling approach for material removal at the abrasive scale, based on copper oxidation rates influenced by intermittent mechanical action.

MECHANISTIC MODEL FOR COPPER CMP

As discussed above, passive or protective films form on copper under typical CMP conditions. As these thicken, the oxidation rate of copper decreases at rates that vary for different slurries The films are removed periodically during polishing by interaction with abrasive particles and pad asperities, causing a dramatic increase in oxidation rate, followed by progressive thickening of the new passive film and a concurrent decrease in oxidation kinetics.

To capture this mechanism in a mathematical model for the local copper material removal rate during CMP, one must consider: the passivation kinetics of copper in the particular CMP slurry; the frequency and force of interaction of abrasive particles and/or asperities at that particular point on the surface; and the mechanical response of the passive film to forces applied on a sliding abrasive/asperity. If the passivation kinetics, frequency of interaction, and amount of passive film removed per interaction are known and remain unchanged during the process, and the process operates in a quasi-steady-state (after each interaction the oxidation rate returns to the same value), the removal rate during the process can be obtained as follows (Figure 1).

Let $i(t')$ be the transient passive current density at time t' after bare copper is exposed to a given oxidizing passivating environment, and i_0 be the current density immediately after an abrasive-copper interaction (which would only be $i(t')$ if the interaction removed the entire film). If τ is the interval of time between two consecutive abrasive-copper interactions, and t the time since the last abrasive-copper interaction, with t_0 defined such that $i(t'=t_0)=i_0$, then the average removal rate of copper (in nm/s) between the two abrasive-copper interactions is:

$$\dot{V}_{CW} = \frac{M_{Cu}}{\rho n F \tau} \int_0^\tau i(t_0 + t)dt \tag{1}$$

where M_{Cu} is the atomic mass of copper, ρ is the density of copper, n is the oxidation state of the oxidized copper, and F is Faraday's constant. The integral gives the total oxidizing charge passed during interval τ. Although the oxidation conditions may vary from one abrasive-copper interaction to the next, this is unlikely to be a significant factor for modeling CMP with a continuous flow of slurry and good agitation.

For most commercial pads, τ, t_0 and i_0 would be stochastic variables; the interaction frequency, the duration and force of contact would vary from one abrasive-copper interaction to

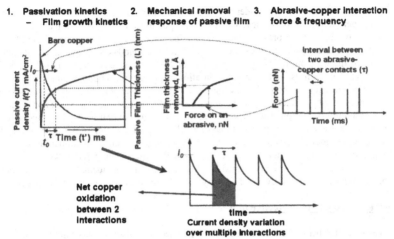

1. **Passivation kinetics**
   – **Film growth kinetics**
2. **Mechanical removal response of passive film**
3. **Abrasive-copper interaction force & frequency**

Figure 1: Determination of copper removal rate during CMP from passivation kinetics of copper, mechanical response of passive films, and abrasive-copper interaction force and frequency

another. Using averaged values of stochastic parameters to evaluate a non-linear function of these parameters, as is the case for copper CMP, could introduce significant errors. A Monte-Carlo based scheme would be appropriate for accounting for these stochastic variations during copper CMP. For pads with well-defined structures, such as fixed abrasive pads, τ, t_0 and i_0 would be expected to be constants that could be determined experimentally.

INPUT PARAMETERS FOR CMP MODEL

To evaluate the CMP model, it is necessary to measure or estimate the parameters within equation 1, and assess the validity of the quasi-steady state assumption underlying equation 1.

Frequency of Mechanical Interactions

The frequency and amount of passive/inhibitor film removed from copper during CMP depends on the frequency, force and duration of abrasive-copper and pad asperity-copper interactions. The pad properties, applied pressure and conditioning determine the size and shape of local contact areas and their spatial distribution. Elmufdi and Muldowney[4] have measured the real contact area of asperities on a typical commercial pad using confocal reflectance interference contrast microscopy (C-RICM), and found the real contact ratio, $Ar_\%$, to be between 1 and 10% for the usual operating CMP pressures. Under conditions where $Ar_\%$ was 0.01, the average asperity contact area, $\overline{Asp_{area}}$ was about 100 μm^2. Taking the relative pad-wafer velocity, V, as 1 m/s, this gives the average interval between consecutive asperity-copper contacts, τ, and the duration of contacts as:

$$\tau = \sqrt{\overline{Asp_{area}}}\Big/ V \cdot Ar_\% = 1 \text{ ms} \qquad \text{and duration of contact} = \sqrt{\overline{Asp_{area}}}\Big/ V = 10\mu s$$

The distribution of abrasives under each asperity-copper contact is needed to calculate the abrasive-copper interaction frequency. This depends upon the slurry composition and colloidal properties of the abrasive. At present, there are no experimental data on the distribution of abrasive particles under pad asperities. However, we can set bounds on the abrasive interaction frequencies. If there are multiple abrasive particles under the same asperity, then the interval between these contacts must be less than the duration of the asperity-copper contact, i.e. the abrasive-copper interaction interval is less than 10μs. If there is an abrasive contact once every (or every few) asperity contact(s), then the abrasive interaction interval will be similar to the asperity contact interval, about 1ms. Regardless, since the time interval between consecutive asperity contacts is about 2 to 3 orders of magnitude larger than the interval between consecutive abrasives contacts under the same asperity, the electrochemical changes on copper between sequential abrasive contacts under the same asperity will be minor compared to the electrochemical changes between two asperity contacts. This justifies using a single parameter to describe passive film removed by a pad asperity and all abrasive particles under that asperity.

Mechanical response of protective film

To evaluate the tribo-chemical model of copper CMP one needs to know how much material is removed by each interaction between the passivated copper surface and a pad asperity. Typical copper removal rates during CMP are in the range of 50 to 600 nm/min. For intervals between two asperity copper contacts of 1 to 10ms, this corresponds to removal of a copper layer of 0.1 to 1Å thick per interaction (due to both dissolution between the two interactions and removal of oxidized copper film by the interaction). This is less than 1 atomic layer (the atomic radius of copper is 1.4Å). Since one cannot physically remove a fraction of an atom, this means that the likelihood of removal of a single surface copper species is less than unity per interaction. This is consistent with experimental observation that the roughness achieved during copper CMP is less than a nanometer. The mechanical phenomena during CMP must be akin to the plucking of certain atoms/molecules from the surface during each asperity/ abrasive interaction (the "chemical tooth" model proposed by Cook[5]). The likelihood of removal would depend upon the chemical environment, which determines the affinity of the surface species for the abrasive particles/pad asperities. This seems a more reasonable explanation for the observed synergy between chemistry and mechanical effects in CMP than the generally accepted theory that the chemical environment affects the mechanical properties of surface layers.

Passivation Rates and Quasi-Steady State Condition

Figure 2 shows schematically the attainment of steady state during copper CMP. The oxidation current decreases with time as protective films of oxide or inhibitor develop, while the thickness (or completeness) of these films increases. CMP is assumed to start at time t_{0b} ('b' denotes before abrasion), where the passive film thickness is 0_b. Part of the film is removed (denoted by the vertical arrow), leaving a film thickness of 0_a. The film grows to 1_b during the interval τ, when more of the film is removed. The process continues, eventually reaching a quasi-steady state, where the amount of film formed between abrasions is equal to that removed by the abrasion.

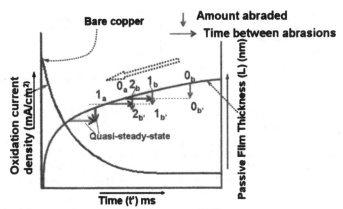

Figure 2: Attaining a quasi-steady-state during copper CMP.

The functional form of the current decay shown schematically in Figure 2 was obtained from potential-step experiments using a copper microelectrode[6] and is taken as:

$$i(t) = i_{bare} \cdot \left(\frac{t}{t_{bare}}\right)^{-n} ; \forall t \geq t_{bare} \qquad (2)$$

where i_{bare} is the current density on bare copper, up to the point when the formation of oxidized films or adsorption of inhibitor causes current decay. The integral of the current with respect to time gives the total material removal (through Faraday's law). However, only a portion of this is responsible for the thickening protective film; the rest represents copper that dissolves in the slurry.

PREDICTIONS

The charge of oxidized copper in a passive film (a convenient measure of the thickness) was estimated using representative literature values for model parameters, assuming different passive film thicknesses at the time of the first abrasion. These film thicknesses are shown in Figure 3 in terms of the time that the copper was exposed to the slurry before polishing started. With short initial exposure times, the film must thicken before attaining quasi-steady state, whereas with long initial exposure times it must thin down. It may take on the order of 10s before quasi-steady state is attained, but this is still relatively short compared to the duration of CMP (of the order of 100s). Hence a model based upon quasi-steady state seems reasonable.

The material removal rate is the primary parameter of interest from CMP models. We are currently working on detailed evaluation of the amount of dissolution occurring as a function of the thickening film, using potential-step chronoamperometry data. As a first approximation, however, Figure 4 shows the predictions from the integrated tribo-chemical model, assuming that 50% of the current passing at any time results in copper ions dissolving directly into the slurry. One sees that if the original protective film was very thin, there is rapid material removal in the first few seconds of polishing, until quasi-steady state is reached. In contrast, if the film is thick, the removal rate is lower for several seconds until quasi-steady state is reached.

39

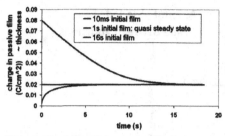

Figure 3: Simulation of charge of oxidized copper in passive film for different initial film formation times: 1ms interaction frequency (τ); $10\mu C/cm^2$ charge in thickness removed per interaction; passivation kinetics – $i(t)=0.01t^{-0.5}A/cm^2$ and $t_{bare}=0.1ms$

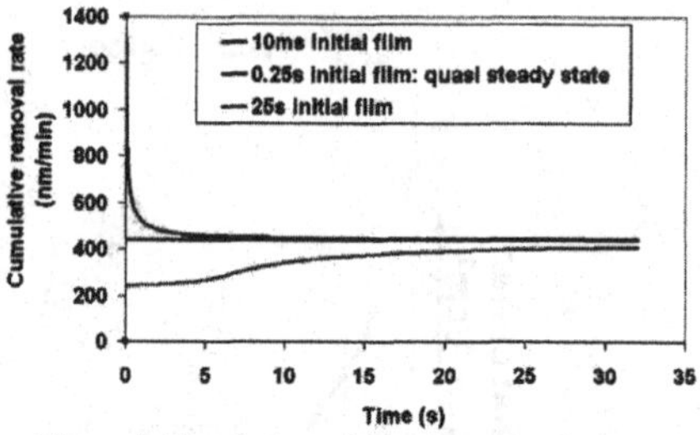

Figure 4: Simulation of copper removal rate over time for different initial film formation times, assuming 50% direct dissolution of oxidized copper, other parameters as for Figure 3

CONCLUSIONS

A new tribo-chemical copper CMP model is based on the physical mechanism responsible for the synergism of mechanical and chemical effects during CMP, unaddressed in previous models. Copper removal is tracked by the oxidation rate of copper, which fluctuates significantly due to regular removal of protective species by pad asperities and abrasives in the CMP slurry.

ACKNOWLEDGMENTS

This work was funded by AMD, Applied Materials, ASML, Cadence, Canon, Ebara, Hitachi, IBM, Intel, KLA-Tencor, Magma, Marvell, Mentor Graphics, Novellus, Panoramic, SanDisk, Spansion, Synopsys, Tokyo Electron Limited, and Xilinx, with donations from Photronics, Toppan, and matching support by the U.C. Discovery Program.

REFERENCES

1. T.C. Zhang, X.X. Jiang, S.Z. Li, "Acceleration of corrosive wear of duplex stainless steel by chloride in 69% H_3PO_4 solution" *Wear* **199** (1996). 253–259.
2. Y. Ein-Eli, D. Starosvetsky, "Review on copper chemical–mechanical polishing (CMP) and post-CMP cleaning in ultra large system integrated (ULSI)—An electrochemical perspective," Electrochimica Acta, **52**, (2007) 1825-1838
3. G. Xu, H. Liang, J. Zhao, Y. Li, "Investigation of Copper Removal Mechanisms during CMP," Journal of The Electrochemical Society, 151 (10) G688-G692, (2004)
4. C.L. Elmufdi, G.P. Muldowney, "A Novel Optical Technique to Measure Pad-Wafer Contact Area in Chemical Mechanical Planarization" Mater. Res. Soc. Symp. Proc. V91, 2006 Spring
5. L.M. Cook, "Chemical processes in glass polishing," J. Non-Cryst. Solids 120, 152 (1990).
6. S. Tripathi, "Tribochemical Mechanisms of Copper Chemical Mechanical Planarization (CMP) – Fundamental Investigations and Integrated Modeling", Ph.D. Dissertation, University of California, Berkeley, December 2008.

CMP of Emerging Materials

Mater. Res. Soc. Symp. Proc. Vol. 1157 © 2009 Materials Research Society   1157-E03-02

Chemical- Mechanical Polishing of Optical Glasses

Elisabeth Becker, Andreas Prange, Reinhart Conradt
Institute of Mineral Engineering, Department of Glass and Ceramic Composites, Mauerstrasse 5,
D-52074 Aachen, Germany

ABSTRACT

This paper presents the Chemical-Mechanical Polishing of optical and fine-optical glasses, which is employed to fulfill the optical requirements for the surfaces of optical lenses. We present the effect of chemical interactions in the polishing process of optical lenses and show how these interactions can be influenced by the additions of certain chemicals. Furthermore, we present a thermodynamic simulation tool, by which these interactions can be modeled. Thus it is possible to understand the chemistry in the suspension and to simulate recent polishing processes in advance, with new additives or with various intrinsic glass ions from different glasses.

INTRODUCTION

The final step in the production of optical lenses is the polishing process, which is employed to bring about the desired surface quality (minimum roughness, absence of surface and sub-surface flaws) as well as the final adjustment to the macroscopic shape for optical applications. For both objectives, a certain removal of glass material is necessary. This production step is very critical because it is time and cost consuming. For our experiments we used the so called Synchro Speed Process: A rotating lens is polished by an also rotating polishing pad with a polishing suspension of water and polishing grain (CeO_2). After the optimization of the mechanics of the polishing machine [1], the importance of chemical parameters became obvious. Therefore we focused on the polishing grain substrate [2], [3] and its behavior as a colloidal system [4] and on the interaction between the glasses, a multiplicity of oxides and non-oxides, and the aqueous system [5]. As an important factor, the polishing suspension is reused for polishing either the same glass or different glasses. In the first case, the suspension will be concentrated with intrinsic ions; in the second case, a mixture of ions from different glasses will be found in the suspension. The kind and amount of different ions influences the colloidal chemistry of the polishing suspension, its stability and the agglomeration state of the polishing grains. These interactions between the glass specific ions can cause instabilities in the polishing process yielding a decrease of the glass removal or scratches at the glass surface.

EXPERIMENTS

In several experiments, it was found that a mixture of glass ions can lead to instabilities in the polishing process. For example, the equilibria between Si, Ca and Zn ions in aqueous solutions are very sensitive and a shift of these equilibria can cause big differences in the removal rates. This is demonstrated in Figure 1, which shows the concentration of the diluted species (left ordinate) and the resulting removal rate (right ordinate) as a function of the polishing time for three different suspensions; suspension 1 and 2 are randomly sampled suspensions used in industrie, suspension 3 is a fresh suspension prepared for a laboratory scale experiment. The glass polished in all three cases is BaK 4. This is a typical optical glass, causing few problems in the polishing process. Therefore it is used as reference glass. The left diagram shows that a well pre-conditioned suspension yielding good polishing results requires a certain amount of dissolved matter, which corresponds to a specific solubility product depending on the glass to be polished. The stability of the suspension is reflected by the constant level of elemental concentration. The concentration levels remain constant although the input of glass removal into the aqueous system increases steadily. This presumes the temporally constant occurrence of precipitates which control the concentration levels via their solubility products. Note: The boron level stems from the previous removal of a different glass obviously compatible with BaK 4. Suspension no. 2 is a suspension preconditioned by polishing a lead silica glass. With respect to BaK 4, the Si and Ba levels are much too high. In other words, the input of BaK 4 removal leads to a destabilization of the suspension; hence the very low removal rate. Finally, the third diagram shows what happens during the pre-conditioning process of a suspension. We started with an aqueous system containing a high level of Ca^{2+} ions, but no glass removal yet. The course of the Si concentration shows an increase proportional to the removal followed by the formation of a sequence of precipitants, presumably $ZnSiO_4$ at the first kink, and a Ca silicate at the second kink. The occurrence of the Ca silicate goes together with a dramatic drop of the removal rate, also reflected by a drop of the power uptake of the polishing machine (not documented here). Even after 80 min of preconditioning, the concentrations typical of BaK 4 and along with this, temporally constant removal rates are not reached yet.

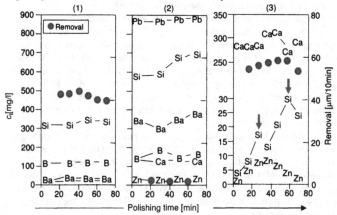

Figure 1. Correlation between chemical equilibria in the suspension and removal rate in the polishing process; polished glass: BaK 4; (1) an industrial suspension well conditioned for BaK 4, (2) an industrial suspension preconditioned for a lead silicate glass, (3) a fresh lab suspension with $CaCl_2$ addition

Another example is an unacceptably high roughness which is initiated by the critical presence of fluoride and alkaline earths in the polishing suspension. It was assumed that the formation of alkaline earth fluorides causes this high roughness, as they function as rogue particles in the suspension [6].

The two above examples clearly show that the polishing process of optical lenses strongly depends on the glass composition and on the interaction between different kinds of glasses which are polished consecutively. To classify this dependency systematically, the discussed glasses were divided into "glass families" (Table I), depending on the composition and on their polishing behavior [7]. Polishing glasses from one glass family in a mixed sequence does not cause any problems. Family 1 contains the reference glasses, BaK 4 and BK 7 (SiO_2-rich) known to cause few problems only.

Table I. Optical glasses, combined to different compositional glass families

Family	Glass	Composition
1	BaK 4	SiO_2-rich
	BK 7	
2	KG 1	SiO_2-rich,
	N-LaK 8	CaO-containing
2a	N-FK 51	like 2, but fluoride glass.
3	SF 6	SiO_2-rich,
	N-SK 16	CaO-poor

In addition to these experiments, different additives were added to the polishing suspension to systematically influence the chemical equilibria in the polishing suspension and therefore the polishing result [8]. Basim et al. showed that this procedure works well for silica polishing suspensions [9]. Here we will show the results of the polishing experiments with additions of citric acid, which had an especially high influence on the removal rate. Figure 2 shows the removal of the polishing process and the roughness of the polished glass without and with the addition of different amounts of citric acid. The addition of 0.5 g/l does not affect the removal very much, but the addition of 2.0 g/l citric acid doubles the removal rate for nearly all glasses investigated. At the same time, the roughness of the product increases significantly for the small amount of citric acid. With a sufficient amount of the acid, the roughness is not impaired anymore. Thus, by finding the right additive in the right concentration the removal per time can be doubled without any losses in the surface quality.

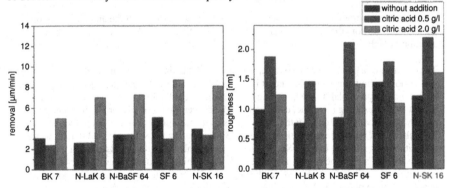

Figure 2. Removal and roughness depending on the amount of the additive citric acid

SIMULATION OF CHEMICAL PROCESSES IN AQUEOUS SOLUTION

To examine the complex chemical processes in the aqueous solution (polishing suspension), the polishing process is modeled with the software HSC Chemistry from Outokumpu [10]. For this simulation it was assumed that in the polishing process one thin layer of glass after the other will be set into thermodynamical equilibrium with the aqueous system, see [11]. Thus a reaction path for the interaction of the individual glass constituents in the polishing suspension is obtained. The calculation starts with 55.51 mol H_2O (corresponding to 1 kg H_2O, or to 1 l H_2O in good approximation). Then elements are introduced into the system in steps of 1 µg glass in proportion to the glass composition under consideration. In certain applications, a specific amount of salt is added to the water prior to dissolving the glass.

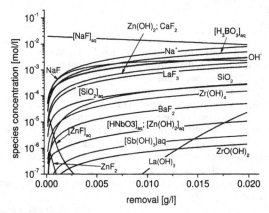

Figure 3. Concentration of species in aqueous solution as a function of the amount of dissolved glass (removed from the polished glass) for a typical optical glass N-LaK 8 with the prior addition of NaF

The example of a polishing process where fluoride and alkaline earth are present in the suspension is shown in Figure 3. Here it is the SiO_2-rich and CaO-containing glass N-LaK 8 with the prior addition of NaF to the polishing suspension. The reliability of these calculations can be shown by the resulting pH value, calculated by the concentration of OH⁻ or H⁺. It produces exactly the pH value as found in the experiments (with each glass assuming an individual steady state pH value [5]). In the simulation in Figure 3 the resulting pH value 11.5 is higher than the one without the addition of NaF. So, the changes in the pH value due to additives in the suspension can also be observed. The excess of the solubility product directly causes a deposition of CaF_2, LaF_3, NaF, BaF_2 and ZnF_2. In earlier polishing experiments, it was observed, that the addition of NaF to the polishing process of this glass causes a high surface roughness, which means poor surface quality. It was assumed and now proved, that this is due to the precipitation of several fluorides, especially of CaF_2. As known from Kaller [12] and Sabia and Stevens [13], the best polishing grains are softer than the polished surface and do not mechanically remove material. Though the fluorides are not the polishing grain, they can damage the glass surface by scratching, as they form hard, insoluble crystals. This assumption is supported by the practice in the optical technology, where fluoride glasses are often handled in a particular way, even without this knowledge. Other deposits calculated are SiO_2, $Zn(OH)_2$, $Zr(OH)_4$, $La(OH)_3$ and $ZrO(OH)_2$. Here it must be considered that the calculation gives the thermodynamically stable state only. In contrast to simple ionic salt, the occurrence of precipitates of the salt hydroxides and oxides require a certain oversaturation (often several times the equilibrium concentration). The calculations only give the thermodynamic threshold for precipitation. But this threshold has much practical significance, too. Once precipitated, a species can only be dissolved again by reducing the concentration under this critical value. As mentioned earlier, the occurrence of a precipitate can cause instabilities in the polishing process, like reducing the removal rate or impact on the surface quality.

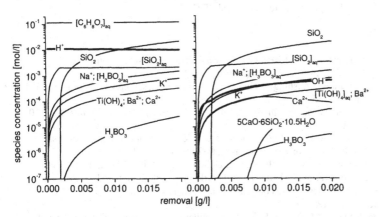

Figure 4. Concentration of species in aqueous solution as a function of the amount of dissolved glass (removed from the polished glass) for the reference glass BK 7 with and without the addition of acidic acid ($C_6H_8O_7$)

As shown in the experimental part, citric acid has a strong influence on the results of the polishing process. To check the chemical impact of this addition, we present the simulation of the polishing process of the reference glass BK 7 with and without the addition of citric acid (Figure 4). It can easily be seen that the pH value changes because of the addition of the acid and at the same time, the precipitate $5CaO \cdot 6SiO_2 \cdot 10.5H_2O$ cannot appear any more. Citric acid has this effect of inhibiting certain precipitates for all glasses analyzed, which explains its positive effect on the polishing process.

CONCLUSION

The chemical state of the polishing suspension has an obvious impact on the polishing process, namely on the removal in the process and the quality of the polished surface. It was shown that a constant species concentrations in the polishing suspension yields high removal rates, while temporally strong changes cause a drop of the removal rate. These changes take place during the preconditioning of the polishing suspension or they are evoked by polishing incompatible glass with the same polishing suspension. The results demonstrate that crossing a new solubility limit provokes a potential failure in the process. The chemical state of the suspension can be systematically improved by choosing suitable chemical additives. This was demonstrated for citric acid. The kind of chemical interaction in the suspension can be explained and predicted by thermodynamic modeling.

ACKNOWLEDGMENTS

The research project (AiF-Nr. 15763 N) was financed within the agenda for promotion of industrial collective research (IGF) by the Bundesministerium für Wirtschaft und Technologie via the Arbeitsgemeinschaft industrieller Forschungsvereinigung (AiF) and the Forschungsvereinigung Feinmechanik, Optik und Medizintechnik (F.O.M.).

REFERENCES

1. S. Hambücker: Steigerung der Effizienz bei der mechano-chemischen Politur von Glaswerkstoffen, Glas-Ingenieur No. 2 (2003); p. 38-44
2. T. Izumitani, S. Harada: Polishing mechanism of optical glasses, Glass Technologie Vol. 12 No. 5 (1971); p. 131-135
3. L. M. Cook: Chemical Processes in Glass Polishing, J. of Non-Crystalline solids, No. 120 (1990); p. 152-171
4. E. Becker, A. Prange, R. Conradt: Chemical-Mechanical polishing of optical glasses, Proceedings International Congress on Glass (2007); Strasbourg, France
5. R. Conradt, A. Prange, E. Becker, F. Klocke, U. Schneider, O. Dambon: Zetapotential Einfluss des Agglomerationsverhaltens von Polierkörnern, Glasabrieb und verwendetem Poliermittel auf das Polierergebnis optischer Gläser, Report AiF (industrial collective research) Nr. 13582N (2006)
6. T. Suratwala, R Steele, M. D. Feit, L. Wong, P. Miller, J. Menapace, P. Davis: Effect of rogue particles on the sub-surface damage of fused silica during grinding/polishing, Journal of Non-Crystalline Solids 354 (2008), p. 2023-2037.
7. R. Conradt, U. Dahlmann, S.-M. Groß, F. Klocke, S. Hambücker: Optimierung der chemischen Einflüsse bei der mechanischen Politur von Glas, Report AiF (industrial collective research) Nr. 12063N (2001)
8. R. Conradt, A. Prange, E. Becker, F. Klocke, U. Schneider, O. Dambon: Prozesskontrolle bei der Politur optischer Gläser, Report AiF (industrial collective research) Nr. 13582N (2006)
9. B. Basim, B. Moudgil: Slurry Design for Chemical Mechanical Polishing; KONA (2003), no. 21, p. 178-183
10. A. Roine, HSC Chemistry 6.0. Outokumpu Research Oy. Pori, 2006.
11. E. Becker, A. Prange, R. Conradt: Simulation of the chemical influences to the polishing process of optical glasses, Proceedings European Society of Glass (2008); Trencin, Slovakia
12. A. Kaller: On the polishing of glass, particularly the precision polishing of optical surfaces, Glastechnische Berichte 64, Nr. 9 (1991), p. 241-251
13. R. Sabia, H. J. Stevens: Performance Characterization of cerium oxide abrasives for chemical-mechanical polishing of glass; Machining science and technology 4 (2000), no. 2, p. 235-251.

Mater. Res. Soc. Symp. Proc. Vol. 1157 © 2009 Materials Research Society   1157-E03-08

Opportunities and Challenges to Sustainable Manufacturing and CMP

David A. Dornfeld

University of California, Mechanical Engineering, Berkeley, CA 94720-1740, USA

ABSTRACT

Today the requirements for reducing the impact of our manufacturing activities are increasing as the world awakes to and addresses the environmental impacts of our society. Energy consumption, greenhouse gas emissions, materials availability and use, environmental impact levels, etc. are all topics of interest. Semiconductor manufacturing in general and process steps such as CMP are not exempt from this and, in many cases, the industry has led the efforts in reducing impacts. This paper will first review some of the drivers for sustainable manufacturing, then define some of the terms that will be useful for determining the engineering aspects of sustainability and sustainable manufacturing, as well as metrics for assessing the impact of manufacturing in general and CMP in particular. An assessments of CMP will be given to illustrate the potential for "design for the environment" in CMP and related processes. Consideration will be given to research opportunities, including process modeling, that this focus provides to CMP researchers, consumable suppliers and industry.

INTRODUCTION

The move towards sustainable systems and development is accelerating due to the input on the challenges facing the planet from a large number of directions – from scientific societies to consumers. How this movement is perceived by manufacturers and responded to by the engineers engaged in production is not clear. The definitions are varied and, often, not scientifically based. The metrics for comparison or quantification not defined. The relationship between environmental, societal and business aspects of sustainability and engineering principles governing manufacturing are not clear. However, it is important that efforts be made to help define these terms, derive these metrics and help establish these critical relationships. This is important to meet the demands of government, consumers, and the competitive market. From a business perspective, the uncertainty about some of these issues pose real risks to businesses. Finally, as educators, it is important to develop a clear set of engineering principles and tools to equip our students to fully participate in addressing these challenges.

Whether or not one agrees with the predictions about global warming, the generation of greenhouse gases, etc. it is clear that energy and resources of production are costly and the costs are likely to increase. Costs of control and disposal of materials used in production, along with the normal environmental health and safety requirements, are likely to increase as well. Scarcity of water and other materials, independent of environmental concerns, is another risk to industry. Lester Brown underlined the importance of this as follows [1]:

> "Today, we need a shift in how we think about the relationship between the earth and the economy. This shift is no less fundamental than the one proposed by Copernicus back in 1575. This time, the issue is not which celestial sphere revolves around the other -- but whether the environment is part of the economy -- or the economy is part of the environment."

One of the issues is the definition of "sustainable manufacturing." No clear definition exists. We will begin with a short discussion of sustainability in the context of manufacturing and, then, discuss efforts in semiconductor industry, with a specific focus on chemical mechanical planarization (CMP), to "green" the processes.

SUSTAINABLE MANUFACTURING AND GREEN MANUFACTURING

Sustainability is usually used in reference to development (see, for example, the Brundtland Commission report of the UN [2]). Brundtland states that "Sustainable Development is development that meets the needs of the present without compromising the ability of future generations to meet their own needs." It is not obvious how to apply this directly to manufacturing but we could consider several categories of "sustainability." These categories depend on whether or not the resource consumed, or impact, is renewable or the impact can be accommodated by the environment. Materials and other resources extracted from the earth, like coal and oil, are not usually considered to be "renewable". So electricity generated from these resources is not renewable. Solar and wind based energy would be considered renewable. Impacts can be sustainable if the environment can "absorb" the impact as part of the natural processes. Hence, there is a limit to the amount of CO_2 (and other green house gases (GhG)) or other contaminants put into the atmosphere based on the atmosphere's ability to accommodate this input. Exceeding this level is not a sustainable situation. Some resources may appear to be sustainably used but are scarce. For example, water is "renewable" in a specific area if not used beyond the ability of the aquifer or water sources to replenish that taken from the system. So, recycling water used in an industrial process may be a sustainable solution. Of course, one needs to consider the energy and material used in treating the water as part of any recovery system.

The use of "exergy" to assess sustainability of a process or system has been proposed, for example, [3]. Exergy is defined as the measure of the usefulness of value or quality of a form of energy (thermal or chemical) and measures the "maximum amount of work which can be produced by a system or flow of matter or energy as it comes into equilibrium with a reference environment", [3]. It is helpful to identify causes of inefficiencies. But, rather than expecting an impossible efficiency, it compares the system or flow the thermodynamically maximum – for example Carnot efficiency. It reflects the irreversibilities associated with the process or system.

Hence, it seems as if it will be necessary to use a combination of assessment tools and definitions, depending on the specific materials and resources consumed and the impact that consumption has, in the course of the process, on the environment. The subject is still under intense discussion. For the purposes of this paper, we will assume that improvements that reduce energy or material/resource consumption (including water) and/or reduce the impact of the process or system on the environment are "more green" than other processes and systems and are "better." How much better and whether or not the improvement is worth it in terms of the energy or materials/resources consumed (and their impacts) to make the improvement is an outstanding question. This requires engineering metrics to make these assessments and this is part of the challenge.

GREENING OF CMP

The semiconductor industry has been proactive in improving processes and systems to reduce energy and material use and the process impacts. This started with the natural attempts to reduce the cost of ownership in any way possible for processes that use exceptionally expensive materials (solid, liquid and gas) in production either as process chemicals or for cleaning and control. In addition, many of these materials are harmful and environmental health and safety programs have been aggressively developed. Documents such as Semi S23-0705 Guide for Conservation of Energy, Utilities and Materials [4] offer useful guidelines to the industry and address continuous improvements. The ITRS 2005 industry road map anticipates 40% less energy per cm^2 in 2010 from 2005 baseline for Fab tools consumption. But the challenge to the industry is immense.

In the US, in 2004, the semiconductor industry emitted 4.7 Million metric tons of CO_2 eq in global warming gas (GWG) emissions [5]. Based on a global capacity of 2 million wafers per month [6] (and we realize that may be substantially different in today's economic environment) this is the equivalent of approximately 30 million metric tons of CO_2 equivalent of direct emissions annually and approximately 130,000 million kilowatt hours of electricity consumed per year [7].

If we drill down to CMP as part of semiconductor fabrication we see that the actual energy consumption of CMP per wafer is relatively small (less than 0.5%) [6]. This is dwarfed by other processes, like litho, and facilities use. Use of ultrapure water by CMP is quite large (estimated at almost 18% of fab usage) and second only to wafer clean. Technology predictions indicate additional interconnect layers and, hence, additional CMP [7]. But there are a lot of opportunities for improvement, or "greening", of CMP. These are discussed below with respect to means for achieving sustainable manufacturing and, under research opportunities, specific means for improvement.

There are a number of approaches to achieving sustainable manufacturing, or at a minimum, green manufacturing that have applicability to CMP. Allwood [8] lists five basic generic options as:

- Use less material and energy
- Substitute input materials: non-toxic for toxic, renewable for non-renewable
- Reduce unwanted outputs: cleaner production, industrial symbiosis
- Convert outputs to inputs: recycling and all its variants, and

51

- Changed structures of ownership and production: product service systems and supply chain structure

To some extent, all of these are applicable to semiconductor manufacturing in general and CMP in specific. As mentioned earlier, the first option, use less material and energy, has been a driver in process improvement in the semiconductor industry to some extent for a long time. To achieve greener production and, in the extreme, sustainable production, requires reducing the gap between the amount of energy and resources used and the impact of that usage to a more efficient production. Or, as stated in a recent EU study on energy efficiency in manufacturing, moving from maximum gain from minimum capital to maximum value from a minimum of spent resources [9]. This is a logical follow-on from what Lester Brown pointed out earlier. This gap will be filled with a combination of technology "wedges" that, combined, achieve the goal of sustainable production [10]. These wedges consist of specific improvements or modifications in the process or system meeting certain requirements, including that the cost of the materials and manufacturing (in terms of energy consumption and green house gas emissions, etc.) associated with the wedge cannot exceed the savings generated by the implementation of the wedge technology over the life of the process or system in which it is employed. Another requirement is that the cost and impact of the technology must be calculable in terms of the basic metrics of the manufacturing system (that is linking green house gas emission, etc and process metrics like yield, throughput, lead time, etc.).

A key requirement in this "greening" process is the development and application of engineering metrics for ascertaining the life cycle cost of the technology or changes anticipated and evaluating alternatives and different scenarios of implementation. In order to be able to insure that we are moving the process towards a more "environmentally benign" operation (more green and, eventually more sustainable) we need to be able to measure the cost and effectiveness of any changes. This is done with metrics. We don't have room to elaborate on this here but an excellent background on the nature and structure of these metrics is given in [11]. These metrics rely on a clear determination of the goal of the metric (for example, water scarcity or energy independence), metric type (for example, return on investment in green house gas reduction or consumption factor), manufacturing scope (for example, process/tool, line, plant, or supply chain) and geographic scope (for example, local or regional or international). Some common metrics include: Energy payback time, Water (or materials, consumables) payback time, Greenhouse gas return on investment (GROI), and Carbon footprint.

We present in the following section some examples of potential improvements following this discussion of greening CMP.

RESEARCH OPPORTUNITIES

If we review CMP relative to the whole semiconductor manufacturing process we see that, relative to energy, CMP impacts are seen in ultrapure water (UPW) use and associated energy for production and delivery but little impact on GhG "in process" and little influence on energy use overall. CMP consumables not considered in these studies
- abrasive (manufacture and delivery), disposal
- slurry fluids (manufacture and delivery), disposal

- pad manufacture and delivery

These could be important in a complete analysis as is the embedded energy and materials in fabrication, delivery, installation and maintenance of the tool.

We need to first view CMP over the entire scale of its impact, from the meter sized machine and pad to the nanometer sized abrasive particle at the pad/wafer interface. Doing so, we can identify a number of areas for improvement. Following the five basic options for making a process more sustainable outlined above as they apply to CMP, we can find a number of possibilities as:

- Use less material and energy
  - more efficient motors/reduce "idle" time running
  - increase pad life/reduce conditioning/optimize conditioning
  - reduce slurry use
  - more efficient wafer cleaning/reduced use of UPW
  - alternative cleaning methodologies

- Substitute input materials: non-toxic for toxic, renewable for non-renewable
  - alternate slurry chemistry (or non-slurry planarization)
  - greater role of pad topography

- Reduce unwanted outputs: cleaner production, industrial symbiosis
  - fixed abrasive pads to reduce pad loss/slurry waste
  - reduce water use (e.g. during "idle")

- Convert outputs to inputs: recycling and all its variants
  - recycle slurry aggressively (designed for reuse)
  - recover energy from operation (spindle or platen motor)
  - end of life strategies for tool and consumables

- Changed structures of ownership and production: product service systems and supply chain structure
  - lease of CMP tools (i.e. tool manufacturer takes "cradle to grave" responsibility for tool)
  - lease slurry and pads

Many of these areas are actively being explored already. Mapping these possibilities onto the research domain specific research opportunities include:

- Modeling of CMP can contribute to design for manufacturing (DfM) strategies to insure optimum use of the process
- Consumable "design" for extended pad life or reuse
- Defined surface features for optimized contact
- Defined surface features for ideal contact frequency (as in copper CMP)
- Variable compliance (e.g. MEMS-based or electro-rheological actuators)

- "Designer slurries"
- Wafer cleaning to reduce material/water and energy use
- Tool control to reduce set up and tuning with changeover and for idle
- Novel conditioning techniques for extending pad life
- Kinematics for reduced footprint
- Design of tool for "refitting" and scaling to new technology and devices
- Alternate planarization techniques for increased efficiency
- *In situ* metrology/sensors for process control

These are challenging as the process research must track a continuously changing set of requirements as industry follows their roadmap.

CONCLUSIONS

Even if it is not straightforward to define sustainable processes or systems, a sustainability focus offers opportunities for all aspects of CMP. At the least, greening of the CMP process is a benefit. As engineering, we should define the terms before others do it for us! We need engineering metrics for analysis of "trade-offs" and proposed improvements to the process so we are sure that, from a life cycle perspective, there is a net gain. And we need to make sure our focus on energy does not obscure other potential risks associated with resource use, notably water. Some aspects of this work are not clear, for example, how to include the impacts on society. But, there are many excellent research / development topics to pursue by our community. To be successful, this requires collaboration between academics and industry and , for sure, this won't be solved by only one researcher or group.

ACKNOWLEDGMENTS

The assistance and input of researchers Sara Boyd, Corinne Reich-Weiser, Chris Yuan and Teresa Zhang in the preparation of this paper is appreciated. Research in the Laboratory for Manufacturing and Sustainability (LMAS) is supported by IMPACT and UC-Discovery Program, the National Science Foundation as part of SINAM project and the industrial affiliates of the LMAS. Additional information on this topic can be found at lmas.berkeley.edu.

REFERENCES

1. Brown, L., "The Second Coming of Copernicus" The Globalist, 2001.
2. Brundtland Commission Report, UN, 1987.
3. Rosen, M., Dincer, I., and Kanoglu, M., "Role of exergy in increasing efficiency and sustainability and reducing environmental impact," *Energy Policy*, 36, 2008, 128-137.
4. SEMI S23-0705: Guide for Conservation of Energy, Utilities and Materials Used by Semiconductor Manufacturing Equipment, 2005.
5. Inventory of U.S. Greenhouse Gas Emission & Sinks:1990-2004, EPA, April 2006.
6. Boyd, S., LMAS Green Manufacturing Research Presentation, 2008.
7. Cadien, K., "The Future of CMP: Slurries and pads," Keynote presentation, CMP-MIC, Fremont, March 4, 2008.

8. Allwood,J., "What is Sustainable Manufacturing?," Lecture, Cambridge University, February 2005.
9. Anon, "ICT and Energy Efficiency: The Case for Manufacturing," European Commission, Directorate General-Information Society and Media, 2009.
10. Dornfeld, D. and Wright, P., "Technology Wedges for Implementing Green Manufacturing," Trans. North American Manufacturing Research Institute, 2007, vol. 35, pp. 193-200.
11. Reich-Weiser, C., Vijayaraghavan, A. and Dornfeld, D. A., "Metrics for Manufacturing Sustainability," Proc. 2008 IMSEC, ASME, Evanston, IL, October 7-10, 2008.

**Advances in Slurry Particle Mechanism
of Metal and Dielectric CMP**

Mater. Res. Soc. Symp. Proc. Vol. 1157 © 2009 Materials Research Society   1157-E04-07

Accuracy Improvements in LPC Measurements for CMP Slurries

Bruno Tolla and David Boldridge
Cabot Microelectronics Corporation, 870 Commons Drive, Aurora, IL 60504, USA

ABSTRACT

We have examined the Large Particle Count (LPC) analytical method to see whether there are opportunities to improve both the accuracy and precision in hope of improving the utility of the LPC measurement. We have identified weaknesses in the current method that limit both its accuracy and its precision, and which can introduce count errors in excess of a factor of 10. We propose modifications to the current method which result in both accuracy and precision improvements. We recommend these improvements as absolutely necessary for any experiments designed to test the correlation between LPC and defectivity.

INTRODUCTION

Chemical Mechanical Planarization depends on specially designed slurries to produce a flat, defect free surface on a semiconductor wafer in preparation for subsequent processing steps.[1,2] These slurries consist of abrasive particles and active chemistry in a liquid carrier, typically water. To assure effective polishing, the abrasive particle size distribution is tightly controlled. The practitioners of CMP are understandably worried that small quantities of undesirably large abrasive or impurity particles could damage the wafers and lead to yield loss. The quality metric of Large Particle Count (LPC) was implemented to help guard against abnormally high levels of these large particles.

The LPC is presumed to indicate the number of particles larger than a given size, usually stated as 0.56 μm or 1.01 μm. These particles represent only ppm of the mass of the slurry and ppb of the number of abrasive particles, necessitating the use of a highly selective measurement technology. The task of quantifying such a small fraction of the total particle size distribution is quite challenging, and the industry has implemented Single Particle Optical Sizing (SPOS) as a routine analytical technique.[3-8] This technique is nominally capable of detecting only the largest particles while ignoring the smaller, necessary abrasive component. While there is ample evidence that the SPOS technique can provide a warning signal in extreme cases, continued improvements in abrasive and slurry production have dramatically reduced the typical LPC levels. As a result, the correlation between LPC and defectivity has become much less clear.[3,9-13]

Weak correlation between LPC and defectivity can be the result of either inadequacy of the LPC measurement protocol or the appearance of non-LPC related defect mechanisms. We have found compelling evidence that inappropriate calibration and operating practices can severely compromise both the accuracy and the precision of the technique. Bringing the calibration and operating procedures into conformance with best known practice are required if the test is have any hope of recovering some of its former utility.

EXPERIMENTAL

The slurries used in this study are commercial dielectric slurries produced by Cabot Microelectronics Corporation. These slurries are an alkaline dispersion of fumed silica in water. All slurries were shaken briefly to homogenize them prior to sampling, and kept tightly closed between experiments. Unless otherwise stated, all LPC values have been normalized to counts/ml of slurry. Concentrations, where stated, are the concentration of silica abrasive in the instrument at the start of the measurement.

The instrument used was an Accusizer 780A® by Particle Sizing Systems (Santa Barbara, CA) operated in standard autodilution mode. Samples of different sizes were introduced using a Eppendorf Reference® adjustable pipette. Two different experimental methods were employed. The first was a conventional measurement where an aliquot of sample is introduced to the reservoir and measured for 60 seconds while diluting. This measurement protocol is nearly quantitative, with approximately 78% of the sample passing through the sensor during the measurement cycle. The second method involved introducing an aliquot to the reservoir, and making 20 distinct measurements of 6 seconds duration as the sample undergoes dilution. This protocol allows a measurement to be made at a nearly constant particle concentration. Raw data was converted to counts/ml of slurry though multiplication by the "dilution factor" and division by the aliquot volume. Dilution was performed using 18 MΩ-cm deionized water.

Particle size standards were NIST certified polystyrene latex beads purchased from Duke Scientific, Palo Alto, CA.

RESULTS AND DISCUSSION

It should come as no surprise to anyone that failure to properly calibrate an instrument results in the degradation of its accuracy. The SPOS method uses a photo-diode to detect changes in scattered (or transmitted) light and to transform that scattered light intensity into a voltage pulse. The magnitude of the voltage pulse is then compared to an empirical calibration curve to assign a "particle size" corresponding to that light scattering event. The standard calibration provided with the Accusizer® and similar instruments is a general-purpose calibration curve with NIST certified standards spanning the operating range of the instrument. The instruments are provided with proprietary hardware and software to interpolate between calibration points.[14] The general purpose calibration curve has not traditionally included a calibration point at 0.56 um, and may or may not include a point at 1.0 um. Nevertheless, this protocol is capable of providing reasonable accuracy of particle size and count when the entire particle size distribution is within the measurement capability of the instrument. Under these conditions, the manufacturer of the Accusizer® specifies an expected uncertainty of <5% in size and <10% in count.

Application of SPOS to CMP slurries presents a special challenge, however. For CMP slurries, the vast majority of the particles lie below the minimum detection limit of the instrument, with a continuous particle size distribution that increases rapidly as the particle size decreases. As a result, a minor error in calibration will cause the instrument to either count particles which are smaller than the desired minimum or to ignore particles which are larger than the desired minimum. This is shown graphically in Figure 1, where the particle size distribution is given for a standard CMP slurry measured under very dilute conditions (maximum particle count rate of ~50/sec). The integration of the shaded area represents particles that would be erroneously included in the LPC by a 5% error in the size calibration. Under dilute measurement conditions, this error is ~2% per nm, corresponding to ~56% count error in the case of a 5% error in size calibration. One calibration curve was found to be in error by 27 nm, representing an estimated count error of 54%. In addition to the absolute error in the measurement, day-to-day variations of approximately 10 nm (corresponding to ~20% error in count) were common, with occasional variations of that magnitude within a single day.

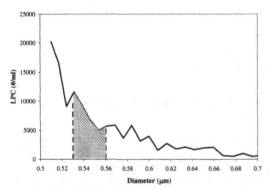

Figure 1: particle size distribution for the "large particle tail" of a CMP slurry. Integration of the shaded area represents particles included in the LPC due to improper calibration.

The calibration problem is compounded by an under-recognized experimental artifact. Traditional CMP slurries have enormous numbers of particles in the desired size range, with estimates of up to ~10^{14}/ml for high-solids concentration slurries. While the SPOS technology will ignore a constant background, random fluctuations in the number of small particles will result in spurious counts by a process known as "secondary coincidence".[15] The effect of secondary coincidence can be seen in Figure 2, where the differential particle size distribution for the same lot of slurry is presented at two different concentrations, measured using the 6 second run duration protocol. The "high" concentration sample (462 ppm abrasive) has many "particles" between 0.56 um and 0.65 um which are not observed in the "low" (177 ppm) concentration sample. For this particular sample, the reported count at the "high" concentration is almost 18 times as large as that reported for the "low" concentration sample, even though both have been normalized to represent the LPC per ml in the original slurry. This discrepancy occurs in spite of the fact that the "high" concentration sample has a maximum count rate of ~3000 counts/sec, well within the manufacturer's recommendation of <9000 counts/sec. It should be noted that the shape of the particle size distribution and the count/ml of slurry become approximately constant below a concentration of ~200 ppm of abrasive, suggesting that the low concentration limit is indeed representative of real particles.

This appearance of secondary coincidence has a profound impact on the error resulting from any miscalibration of the instrument. The increase in slope near 0.56 um causes the count sensitivity to become ~5% per nm of calibration error, suggesting that a calibration error of 5% could result in 140% count error, in addition to the error produced by the "creation" of particles by secondary coincidence.

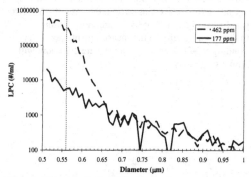

Figure 2: Comparison of particle size distributions determined for high (462 ppm) and low (177 ppm) concentration samples of the same sample of CMP slurry.

A less obvious discovery is a dramatic effect of concentration on precision. Figure 3 shows the scatter in replicate LPC measurements using the traditional (high concentration) LPC measurement protocol. These data represent two production samples, one of which had been declared off-quality and discarded. For these two slurries, random variation produces a relative standard deviation of ~27% of the signal. Extensive signal averaging allows us to estimate that the actual difference in LPC reading is ~17%, suggesting that 44 replicates of each sample would be necessary to achieve 90% confidence limits for both alpha and beta errors. Given the usual measurement protocol of duplicate or triplicate measurement, we estimate that either alpha or beta error will occur ~50% of the time.

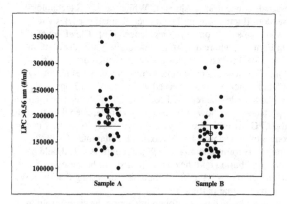

Figure 3: Scatterplot showing variation of LPC reading due to random events. Error bars represent 95% confidence limits on the mean.

Fortunately, precision is also found to be a strong function of concentration. Figure 4 shows the coefficient of variation of one sample measured over a range of concentrations. There is a pronounced maximum in the relative uncertainty at ~300 ppm abrasive, with a decrease at either lower or higher values. The decrease at higher concentrations is the result of the rapid increase in the number of "particles" counted due to the onset of secondary coincidence, while the decrease at lower concentrations appears to represent a true improvement in precision.

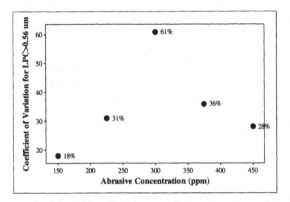

Figure 4: Coefficient of variance as a function of abrasive concentration.

In order to address these observations, we have introduced three modifications to the traditional measurement protocol. We have modified the traditional calibration curve to include a calibration standard at 0.565 um and 1.02 um. The accuracy of these two calibration points is checked on a daily basis, and the instrument re-calibrated any time there is more than a 2.5% deviation from the nominal value. To minimize the time spent calibrating, the total number of calibration points has been reduced to 5, allowing the software to interpolate appropriately while emphasizing size accuracy at the two critical points. Finally, we have reduced the size of the sample so that the maximum concentration is 150 ppm of abrasive, effectively eliminating errors due to secondary coincidence.

CONCLUSIONS

Two key aspects of the LPC measurement have been determined in this work. The first is to ensure that accurate calibration is performed at the critical sizes. Since LPC data is generally reported as number of particles greater than a particular size, those sizes represent points which must be know with a high degree of accuracy, while other sizes are far less critical. The second conclusion of this work is that secondary coincidence can distort the particle count to such an extreme degree that the LPC due to real particles is essentially undeterminable. A fortuitous discovery is that the precision of the LPC measurement actually improves at lower concentrations and reduced count rates, making it possible to measure LPC under conditions where the absolute count is considerably smaller than was traditionally expected. By eliminating secondary coincidence we were able to eliminate an artifact that could lead to overcounting by a factor of 10 or more, and by implementing accurate calibration we were able to improve the count accuracy by an additional factor of up to 50%. In addition, eliminating secondary coincidence led to an improvement in precision of a factor of at least 3.

For the particular slurries examined here, we find that concentrations of <200 ppm abrasive are necessary to avoid secondary coincidence, and optimum precision is observed at ~150 ppm. These correspond to count rates of less than 100 particles/second, and preferably less than 50 particles/second. We anticipate that the exact optimum values will be dependent on the characteristics of a particular slurry. It is critical, therefore, that each slurry be tested to identify the concentration range which provides constant LPC/ml of slurry, and to verify that the precision under those conditions is acceptable. If compromise is necessary, accuracy must be held to a higher standard than precision

These modifications to the traditional measurement protocol are absolutely required to have any reasonable chance of correlating LPC data with slurry performance. Unfortunately, even with these improvements, we are unable to conclude that the LPC measurement is a useful indicator of suitability for use. It appears that we may need to have a "step change" in the measurement technology to achieve the desired level of LPC utility. Along these lines, we are examining the combination of Field Flow Fractionation (FFF) and SPOS as a way to measure LPC without the interference of the small, desirable abrasive particles.[16]

References

1 P. B. Zantye, A. Kumar and A.K. Sikder, Mat. Sci. Eng. R, **45** 89 (2004).
2 J. M. Steigerwald, S.P. Murarka and R.J. Gutman, *Chemical Mechanical Planarization of Microelectronics Materials,* John Wiley and Sons, New York (1997).
3 L. Anthony, J. Miner, M. NBaker, W. Lai, J. Sowell, A. Maury and Y. Obeng, Electrochem. Soc. Proc., **98-7** 181 (1998).
4 K. Nicholes, R. Singh, D. Grant and M. Litchy, Semicon. Int. **24** 201 (2001).
5 J. P. Bare and T. A. Lemke, Micro **15** 53 (1997).
6 D. F. Nicoli, K. Hasapidis, P. O'Hagan, G. Pokrajac and B. Schade, Am. Lab., **33** 32 (2001).
7 D. F. Nicoli, P.O'Hagan, G.Pokrajac and K. Hasapidis, Am. Lab., **32** 18 (2000).
8 L. H Hanus, S. A. Battafarano and A. R. Wank, Micro **21** 71 (2003).
9 E. E. Remsen, S. Anjur, D. Boldridge, M. Kamiti, S. Li, T. Johns, C. Dowell, J. Kasthurirangan and P. Feeney, J. Electrochem. Soc., **153**, G453 (2006).
10 E. E. Remsen, S. P. Anjur, D. Boldridge, M. Kamiti and S. Li, Mat. Res. Soc. Symp. Proc. **867**, W2.4.1 (2005).
11 M. Stutz, H. Barthel and M. Moinpour, MRS Symposium Proc. **991** 0991-C04-02 (2007).
12 K. Nicholes, M.R. Lichty, E. Hood, W.G. Easter, V.B.Bhethanabotla, L. Cheema and D. Grant, 8th Int. CMP-MIC Conf., 221 (2003).
13 M. Bielmann, U. Mahajan and R.K. Singh, Electrochem. Solid State Lett. **2** 401 (1999).
14 D. Wells and D. F. Nicoli, U.S. Patent 5,835,211, (1998).
15 K. T. Whitby and B.Y. H. Liu, J. Coll. Interface Sci. **25** 537 (1967).
16 S. K. R. Williams, I. Park, E. Remsen and M. Moinpour, MRS Symposium Proc. **991** 0991-C09-01 (2007).

Chemical and Physical Mechanism
of Metal and Dielectric CMP

Mater. Res. Soc. Symp. Proc. Vol. 1157 © 2009 Materials Research Society   1157-E05-02

Influence of Chemical-Mechanical Polishing Process on Time Dependent Dielectric Breakdown Reliability of Cu/Low-k Integration

Yohei Yamada and Nobuhiro Konishi
Micro Device Division, Hitachi, Ltd., 6-16-3, Shinmachi Ome-shi, Tokyo 198-8512, Japan

ABSTRACT

The effects of defects caused by Cu chemical-mechanical polishing (CMP) on time-dependent dielectric breakdown (TDDB) in a damascene structure incorporating a low-k interlevel dielectric layer were investigated experimentally. Comb line capacitor structures were prepared with one of three types of defects (rough Cu surface corrosion, Cu depletion, or crevice corrosion) and stressed at 3.2 to 6.2 MV/cm at 140°C. The first two defects had an insignificant effect on the TDDB characteristics while crevice corrosion at the edges of wires significantly degraded them. Investigation of the effects of Cu oxidation during post-CMP cleaning on the TDDB characteristics revealed that the formation of a non-uniform oxide layer accompanying deionized water rinsing was due to the dissolution of Cu oxide during the post-CMP cleaning process. When a barrier metal slurry containing a soluble inhibitor was used, non-uniform oxide formation on the Cu surfaces during post-CMP cleaning degraded the TDDB characteristics. These results demonstrate the importance of uniform Cu oxidation during post-CMP cleaning for improving the TDDB characteristics.

INTRODUCTION

As integrated circuit device density continues to increase and circuit line widths continue to shrink, a low-dielectric-constant (low-k) interlayer and low-resistivity copper (Cu) metallization in the back end of line interconnects are becoming increasingly important in determining device performance. This interlayer and metallization reduce interconnect resistance capacitance delay, cross-talk noise, and power consumption. However, the long-term reliability of low-k dielectric materials is rapidly becoming a critical technical challenge. Low-k time-dependent dielectric breakdown (TDDB) is one of the most important reliability-related issues faced during development of Cu/low-k structures, because low-k materials generally have lower intrinsic breakdown strength than SiO dielectric materials [1–4]. One factor affecting this reliability is the integrity of the diffusion barrier/low-k interface after chemical-mechanical polishing (CMP). A previous study of the effect of the solution used for post-CMP cleaning on the TDDB characteristics [5] showed that, when a barrier metal slurry with a water-soluble inhibitor is used, the characteristics are significantly affected by the type of solution used.

We have now investigated the effects of the CMP process on the TDDB characteristics. Copper corrosion is a critical factor in the Cu metallization process. Severe corrosion can significantly reduce the wafer yield and even minor corrosion can lead to reliability problems. We first investigated the effects of rough Cu surface corrosion, Cu depletion, and crevice corrosion on the TDDB characteristics. Next, we examined the effect of Cu oxide removal during post-CMP cleaning on the characteristics and found that they were sensitive to the cleaning conditions. Finally, we developed a TDDB degradation model that explains the formation of non-uniform Cu oxide during post CMP cleaning.

EXPERIMENT

Model of TDDB [1,2]

Low-k TDDB is the phenomenon of low-k dielectric breakdown due to a high electric field applied for a certain amount of time. The generally accepted model of dielectric breakdown caused by Cu comprises three steps: (1) Cu ionization and injection of Cu^+ from the anode into the dielectric layer; (2) Cu^+ drift through the dielectric layer; and (3) formation of complete leakage passages and accumulation of Cu^+ near the cathode, which results in higher electron leakage current and subsequent dielectric breakdown.

TDDB lifetime measurement

We used the "E-model" to evaluate the TDDB lifetime of Cu interconnects. The E-model is based on the assumption that the TDDB lifetime is exponentially proportional to the electrical field. TDDB lifetime is determined by first applying a comparatively high voltage between electrodes at a predetermined temperature. The time from voltage application to dielectric breakdown is then plotted against the applied electric field, and the lifetime is extrapolated. Our target is a TDDB lifetime at an electric field intensity of 0.2 MV/cm (normal-use state) of more than of 3×10^8 s (~10 years). We measured the TDDB lifetimes of Cu interconnects at 140°C using a test structure consisting of comb line capacitors (L/S = 0.14/0.14 um), as shown in Fig. 1. The line-to-line leakage current was measured as a function of bias stress time using a picoammeter (HP4142B).

Evaluation of TDDB dependence on Cu corrosion defects

Copper interconnects were embedded in dielectric films [layers: plasma-enhanced tetraethylorthosilicate (PE-TEOS); SiCN (k = 4.8); SiOC (k = 3.0); PE-SiO capping layer]. A SiO capping layer was used to protect against sputtering damage, and it was polished using Cu CMP. After patterning by lithography and dry etching of the dielectric layer, a tantalum nitride (TaN)/ tantalum (Ta) barrier metal and a Cu seed were sputtered. Then electroplated Cu film was deposited, and Cu CMP was performed, followed by H_2 annealing. Finally, the wafers were passivated using a SiCN/PE-TEOS bilayer dielectric stack. We used a rotary-type polishing system for both the Cu and barrier-metal polishing processes. A commercially available Cu slurry was used to remove the Cu film up to the barrier metal. The conditions for the Cu CMP were a down force of 13.8 kPa, a platen rotational speed of 93 min^{-1}, a head rotational speed of 87 min^{-1}, and a slurry flow rate of 0.2 l/min. We used a low-selective barrier-metal polishing process to remove the barrier metal and to additionally polish the underlayer of the dielectric film. The amount of dielectric film removed was about 30 nm. The SiOC layer was polished at a rate almost the same as that for the cap-SiO film, which changed its surface properties to hydrophilic from hydrophobic [6,7]. The conditions for the barrier-metal CMP were a down force of 13.8 kPa, a platen rotational speed of 83 min^{-1}, a head rotational speed of 81 min^{-1}, and a slurry flow rate of 0.165 l/min.

After polishing the barrier metal, we used one of several types of cleaning solutions for the post-CMP cleaning. The conditions for the cleaning were a PVA (polyvinyl alcohol) brush rotational speed of 670 min^{-1} and a wafer rotational speed of 1500 min^{-1}. We restricted the queue time between the CMP and the diffusion dielectric barrier deposition to avoid Cu particle formation [2,8]. Sematech 854 pattern wafers were inspected for defects. Common defect types

68

associated with Cu/low-k CMP process are illustrated in Table I. We used an automatic wafer pattern inspection system for chip-to-chip reference comparison. Each defect type was classified using a defect image database. The total inspection area was 124 cm^2, and the defect inspection sensitivity was determined using a pixel size of 0.25 um. The barrier metal slurries and post-CMP cleaning solutions listed in Table II were used to independently investigate the relationship between TDDB lifetime and these defects. The acid barrier metal slurry contained benzotriazole (BTA) as a corrosion inhibitor.

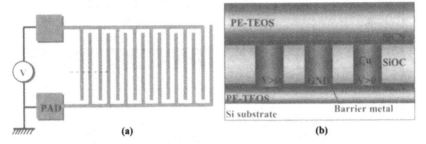

(a)   (b)

Figure 1. Test structure used in this study: (a) comb line capacitor and (b) cross section of structure.

Table I. Corrosion defect types in Cu/Low-k CMP process.

Defect type/ Image			
1	Rough Cu surface	3	Crevice corrosion
2	Cu depletion	4	Pitting corrosion
Impact: Interconnect opens or reliability issues			

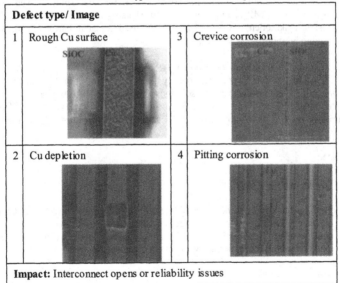

Table II. Slurries and post-CMP cleaning solutions.

Experiment 1: TDDB dependence on corrosion defect		
Copper Slurry	Barrier Metal Slurry	Post-CMP Cleaning Solution

C-1
Abrasive: silica
pH: 3.5
Inhibitor: BTA
	B-1
	Abrasive: silica
	pH: 2.8
	Inhibitor: BTA
		S-1: Organic acid (pH 2)
		    Rough copper surface
		    Crevice corrosion
		S-2: Organic acid salt (pH 3.6)
		    Copper depletion

Experiment 2: TDDB dependence on Cu oxide on Cu surface		
Copper Slurry	Barrier Metal Slurry	Post-CMP Cleaning Solution
C-1	B-2 Abrasive: silica pH: 2.8 Inhibitor: Water soluble	S-1 S-2 S-3: Organic alkali(pH 10.5)

Evaluation of Cu oxide formation during post-CMP cleaning

We prepared three types of cleaning solutions: a citric-acid-based solution, a citric-acid-based salt solution, and a tetramethylammonium-hydroxide (TMAH)-based solution. The oxidized Cu film thickness after post-CMP cleaning was measured using a cyclic voltammeter (CV)(HZ-5000 potentiostat, Hokutodenko Co, Ltd.) [9]. We used an acid barrier metal slurry that contained a soluble inhibitor other than BTA.

Evaluation of Cu surface roughness after post-CMP cleaning

Blanket Cu wafer samples were prepared by polishing using a barrier metal slurry and cleaning using one of the tested cleaning solutions. The Cu surface roughness after post-CMP cleaning was measured with an atomic force microscope (AFM) (E-sweep/SPI4000, SII NanoTechnology Inc.). The roughness of these samples was measured again in acetic acid solution.

DISCUSSION

Corrosion defect effect on TDDB lifetime

As mentioned, we focused on three types of defects: rough Cu surface corrosion, Cu depletion, and crevice corrosion. The density of the surface corrosion was greater than ~ 10 cm^{-2} in the first experiment, and those of Cu depletion were 5.9 and 13.2 cm^{-2} in the second experiment. The measured crevice corrosion defect density was 1.9 cm^{-2} in the third experiment.

Rough Cu surface corrosion

Rough Cu surface corrosion occurs when the entire surface of the metal is exposed to a harsh corrosive environment. In the first experiment, changing the dilution conditions of the citric-acid-based cleaning solutions greatly affected the surface corrosion. The dependence of TDDB lifetime on the corrosion on the CMP surface is shown in Fig. 2. The lifetime negligibly depended on the number of rough Cu surface corrosion defects even though this type of defect reduces device yield.

Cu depletion

Copper depletion expands along a grain boundary when a local region of the Cu oxidized by H_2O_2 is exposed to an acidic environment. In the second experiment, a staging time between the barrier-metal CMP process and the next cleaning step of more than 15 minutes resulted in many Cu depletion defects. The dependence of the TDDB lifetime on the Cu depletion corrosion on the wiring is shown in Fig. 3. The lifetime negligibly depended on the number of Cu depletion corrosion defects even though this type of defect also reduces device yield.

Crevice Corrosion

When localized corrosion occurs below a precipitate on a Cu surface, it is called "crevice corrosion" [10]. This corrosion causes trenches to form at the Cu line edge. In the third experiment, we intentionally corroded the crevices of Cu interconnect surfaces by changing the concentration of the cleaning solution and the rotational speed of the brush used for post-CMP cleaning. The dependence of the TDDB lifetime on crevice corrosion on the CMP surface is shown in Fig. 4a. The lifetime when the defect density was 1.9 cm^{-2} was ten times shorter than when the density was zero. The correlation between the I-V curves and defect density is shown in Fig. 4b. The initial breakdown voltage when the defect density was 1.9 cm^{-2} was lower than when the density was zero. This is because the leakage current when the defect density was 1.9 cm^{-2} was slightly higher. The dependence of the leakage current on the bias stress time at an electrical field strength of 3.2 MV/cm is shown in Fig. 4c. The leakage current in the structure with crevice corrosion was higher than in one without corrosion. Apparently, the breakdown was accelerated by the enhanced Cu ionization at the edges of the wires, where the electric field was concentrated.

Effect of Cu oxide removal during post-CMP cleaning on TDDB characteristics [11]

A comparison of the effects of the CMP processes on the TDDB characteristics of structures subjected to three different types of post-CMP cleaning solutions with a barrier metal slurry containing a soluble inhibitor is shown in Fig. 5. The TDDB lifetime of the structure treated with the organic acid cleaning solution was ten times shorter than that of structure treated with the TMAH-based organic alkali cleaning solution.

The dependence of the leakage current on the bias stress time at an electrical field strength of 3.2 MV/cm is shown in Fig. 6. The sample with the highest leakage current, which was treated with organic acid solution, suffered the earliest dielectric breakdown. The latest dielectric breakdown occurred in the sample treated with the organic alkali solution although the leakage current rose gradually. These results show that the effect of the relatively vulnerable protective film formed on Cu wiring after CMP on line-to-line insulating reliability depends on the dissolution capability of the oxidized Cu in the cleaning solution. Apparently, the change in the form of the Cu wiring surface after post-CMP cleaning, the state of the Cu oxide on the Cu surface, and the amount of dissolved Cu atoms remaining on the SiOC surface between adjacent Cu wires affect the TDDB characteristics.

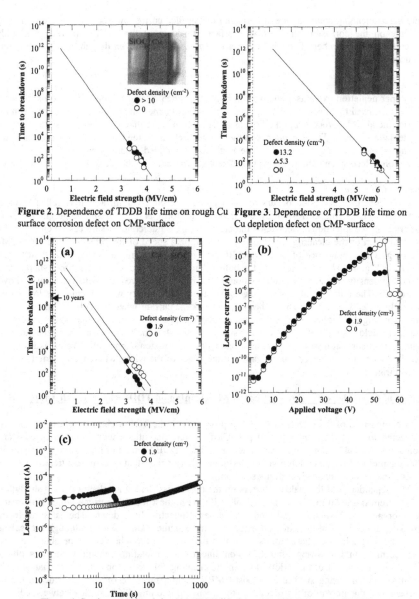

Figure 2. Dependence of TDDB life time on rough Cu surface corrosion defect on CMP-surface

Figure 3. Dependence of TDDB life time on Cu depletion defect on CMP-surface

Figure 4. Crevice corrosion defect: (a) TDDB life time of Cu/SiOC structure, (b) I-V curves measured on comb-wiring pattern, and (c) leakage current on bias stress time.

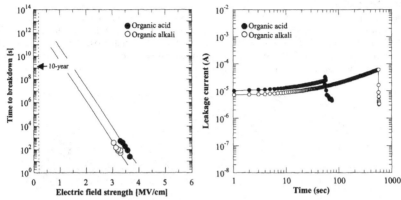

Figure 5. Comparison of TDDB lifetime dependence on post-CMP cleaning solutions.

Figure 6. Dependence of leakage current on bias stress time with different post-CMP cleaning solutions.

State of Cu oxide formed during chemical cleaning

Using a CV, we found no large differences among the samples with similar Cu oxide thicknesses after post-CMP cleaning. In the post-CMP cleaning, a deionized (DI) water rinse process was used after brush cleaning with a cleaning solution. We also measured the Cu oxidation film thickness excluding the effect of the oxidation film formation that occurred during the DI water rinse process. After polished Cu sheets had been immersed in the cleaning solutions (one sheet per solution), we measured the thickness of their cuprous oxide (Cu_2O) film. The current-potential curves for the sheets that were not rinsed with DI water are shown in Fig. 7. Photoelectron spectroscopy analysis showed that each peak was cuprous oxide. The reduction peaks for Cu_2O appeared at a different potential. Because the Cu sheet immersed in a cleaning solution was not rinsed by DI water before the CV measurement, we postulated that the cleaning solution remaining on the sample surface affected the reduction-oxidation reaction during the CV measurement, causing the reduction peak for Cu_2O to shift. At the step before DI water rinsing, the thickness of the cuprous oxide remaining on the surface was different for each sample. Most notably, the cuprous oxide was not etched by the organic alkali solution.

Copper surface roughness after post-CMP cleaning

Figure 8 shows a cross-sectional transmission electron microscopy (TEM) image of a Cu surface after DI water rinsing and of a Cu surface treated with the organic acid salt solution. The surface treated with the salt solution was obviously rougher than that treated with the water rinse. We measured the Cu surface roughness after post-CMP cleaning. The cleaned samples were immersed in an acetic acid solution, which removed the oxidized Cu film completely. Figure 9 shows the Cu surface root-mean-square (RMS) roughness after post-CMP cleaning. The thickness of the Cu_2O layer produced by the polishing process had good uniformity. The RMS roughness of the samples with acidic cleaning after Cu_2O removal was about four to five times greater. The Cu_2O thickness was more uniform after the organic alkali solution cleaning than after the other cleanings. After the organic alkali solution cleaning, the Cu_2O layer formed during the polishing process remained on the Cu surface while the others were produced during the DI water rinse process.

Figure 9 also shows AFM images taken during post-CMP cleaning. We observed many white dots on the Cu surface treated with the organic acid solution following the DI water rinsing. The dots indicate micro projections. Figure 10 shows scanning electron microscopy (SEM) images of Cu surfaces after post-CMP cleaning. Again we observed white dots on the Cu surface treated with the organic acid solution. These dots suggest a non-uniform distribution of Cu_2O.

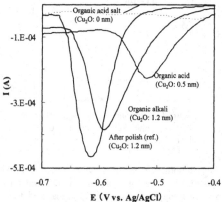

Figure 7. Current-potential curves of CV (no DI water rinse).

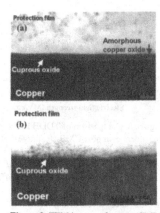

Figure 8. TEM images after post-CMP cleaning(a) DIW rinse and (b) Organic acid solution.

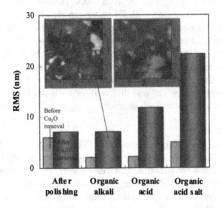

Figure 9. Cu surface RMS roughness after post-CMP cleaning.

Figure 10. SEM images after post-CMP cleaning(a) Organic alkali solution and (b) Organic acid solution.

Cu oxide non-uniform model

From these results, we propose the following mechanism for the formation of Cu_2O film under post-CMP cleaning, which degrades the TDDB characteristics when the barrier metal slurry contains a soluble inhibitor.

Acidic cleaner

The acidic cleaner etches most of the Cu_2O layer produced in the CMP process. The Cu_2O thickness is therefore not uniform after the chemical cleaning. Oxygen penetrates the Cu surface after DI water rinsing. The oxygen concentration varies at the Cu/Cu_2O interface. This difference in the oxygen concentration leads to the formation of discrete anode and cathode regions on the Cu surface, creating an oxygen concentration cell. Figure 11 illustrates Cu dissolution due to the oxygen concentration cell, for which the following reactions might be observed.

In the higher oxygen concentration region at the Cu/Cu_2O interface,

$$O_2 + 2H_2O + 4e^- \rightarrow 4OH^- \tag{1}$$

In the lower oxygen concentration region:

$$Cu \rightarrow Cu^{2+} + 2e^- \tag{2}$$

Alkaline cleaner

The Cu_2O layer remains thick and uniform after chemical cleaning. Oxygen penetrates the Cu_2O layer after DI water rinsing. However, the uniform Cu_2O layer intercepts the oxygen, so the oxygen concentration is very low at the Cu/ Cu_2O interface.

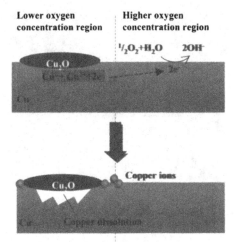

Figure 11. Copper dissolution due to oxygen concentration cell.

TDDB degradation mechanism

Using the results for the Cu surface roughness after post-CMP cleaning, we constructed the model illustrated in Fig. 12 for the TDDB degradation mechanism when using both barrier metal slurry with a soluble inhibitor and an organic acid cleaning solution. Metal ion penetration is a two-step process: emission of metal ions and movement of the ions due to the effects of electrical bias and temperature through the dielectric layer [12]. The Cu ionizes when it comes into contact with the interfacial oxygen and moisture [13]. The oxygen concentration cell due to non-uniform oxidized Cu film created by post-CMP cleaning, plus the applied electric field and temperature during bias-temperature stress, create a driving force for the electrochemical reaction for Cu ionization and migration at the edge of the Cu lines, where the electric field is more intense (Fig. 12a) [14]. Hence, the leakage path along the CMP-surface connects two adjacent Cu lines and subsequently triggers dielectric breakdown (Fig. 12b).

To enhance the TDDB characteristics of Cu/low-k structures, a continuous and uniform oxidation layer must be formed on the Cu surface.

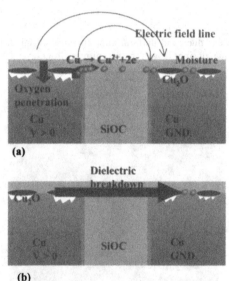

Figure 12. TDDB degradation model.

CONCLUSIONS

We determined that defects, such as crevice corrosion, in the interface between the low-k interlayer and the diffusion barrier shorten the TDDB lifetime. We showed that the TDDB lifetime does not depend on the density of Cu surface defects, such as rough Cu surface corrosion and Cu depletion. We also showed that the cuprous oxide film thickness on the Cu surface after post-CMP cleaning depends little on the cleaning solution. However, we found that the formation of a non-uniform oxide layer that accompanies DI water rinsing is due to the dissolution of Cu oxide during the post-CMP cleaning process. Oxide formation during the post-CMP cleaning degrades the TDDB characteristics when the barrier metal slurry does not contain benzotriazole but does contain a soluble inhibitor. Improvement in TDDB characteristics requires uniform Cu oxidization during post-CMP cleaning in the Cu/low-k damascene process.

ACKNOWLEDGMENTS

We are grateful to Naohito Ogiso, Mutsumi Nakanishi, and Tadakazu Miyazaki of Sanyo Chemical Industries, Ltd. for CV measurements and the fruitful discussion on Cu oxide non-uniform model. We thank Seiko Ishihara, Kiyomi Katsuyama, and Shoji Asaka of Hitachi, Ltd. for their technical support in the investigation of Cu_2O on Cu surfaces.

REFERENCES

1. K. Takeda, K. Hinode, I. Oodake, N. Oohashi, and H. Yamaguchi, in International Reliability Physics Symposium, IEEE, pp. 36–41 (1998).
2. J. Noguchi, IEEE Trans. Electron Devices 52, pp. 1743–1750 (2005).
3. Zs. Tokei, V. Sutcliffe, S. Demuynck, F. Iacopi, P. Roussel, G. P. Beyer, R. J. O. M. Hoofman, and K. Maex, in International Reliability Physics Symposium, IEEE, pp. 326–332 (2004).
4. Y. Yamada, N. Konishi, J. Noguchi, T. Jimbo, S. Kurokawa, and T. Doi, Jpn. Appl. Phys., 47, 6, pp. 4469–4474 (2008).
5. N. Konishi, Y. Yamada, J. Noguchi, U. Tanaka, T. Jimbo, and O. Inoue, in International Interconnect Technology Conference, IEEE, pp. 123–125 (2005).
6. N. Konishi, Y. Yamada, J. Noguchi, and U. Tanaka, in Advanced Metallization Conference, UC Berkeley Extension, pp. 127–132 (2003).
7. Y. Yamada, N. Konishi, S. Watanabe, J. Noguchi, and U. Tanaka, in Chemical-Mechanical Planarization for ULSI Multilevel Interconnection Conference, IMIC, pp. 28–34 (2004).
8. J. Noguchi, N. Konishi, and Y. Yamada, IEEE Trans. Electron Devices 52, pp. 934–941 (2005).
9. S. Nakayama, A. Kimura, M. Shibata, S. Kuwabata, and T. Osaki, J. Electrochem. Soc., 148, pp. B467–472 (2001).
10. A. E. Miller, P. B. Fischer, A. D. Feller, and K. C. Cadien, in International Interconnect Technology Conference, IEEE, pp. 143–145 (2001).
11. Y. Yamada, Y. Yagi, N. Konishi, N. Ogiso, K. Katsuyama, S. Asaka, J. Noguchi, and T. Miyazaki, J. Electrochem. Soc. 155, pp. H301–306 (2008).
12. F. Chen, K. Chanda, I. Gill, M. Angyal, J. Demarest, T. Sullivan, R. Kontra, M. Shinosky, J. Li, L. Economikos, M. Hoinkis, S. Lane, D. McHerron, M. Inohara, S. Boettcher, D. Dunn, M.

Fukasawa, B.C. Zhang, K. Ida, T. Ema, G. Lembach, K. Kumar, Y. Lin, H. Maynard, K. Urata, T. Bolom, K.Inoue, J. Smith, Y. Ishikawa, M. Naujok, P. Ong, A. Sakamoto, D. Hunt, and J. Aitken, in International Reliability Physics Symposium, IEEE, pp. 501–507 (2005).
13. J. Michelon and R. J. O. M. Hoofman, IEEE Trans. Device and Materials Reliability, 6, pp. 169–174 (2006).
14. J. Noguchi, M. Kubo, R. Tsuneda, K. Takeda, N. Miura, and K. Makabe, Jpn. J. Appl. Phys. 44, pp. 94–101 (2005).

Mater. Res. Soc. Symp. Proc. Vol. 1157 © 2009 Materials Research Society   1157-E06-02

Fundamental Mechanisms of Copper CMP –
Passivation Kinetics of Copper in CMP Slurry Constituents

Shantanu Tripathi[1], Fiona M. Doyle[2], and David A. Dornfeld[1]
[1]Department of Mechanical Engineering, University of California,
Berkeley, CA 94720-1740, U.S.A.
[2]Department of Materials Science and Engineering, University of California,
Berkeley, CA 94720-1760, U.S.A.

ABSTRACT

During copper CMP, abrasives and asperities interact with the copper at the nano-scale, partially removing protective films. The local Cu oxidation rate increases, then decays with time as the protective film reforms. In order to estimate the copper removal rate and other Cu-CMP output parameters with a mechanistic model, the passivation kinetics of Cu, i.e. the decay of the oxidation current with time after an abrasive/copper interaction, are needed. For the first time in studying Cu-CMP, microelectrodes were used to reduce interference from capacitive charging, IR drops and low diffusion limited currents, problems typical with traditional macroelectrodes. Electrochemical impedance spectroscopy (EIS) was used to obtain the equivalent circuit elements associated with different electrochemical phenomena (capacitive, kinetics, diffusion etc.) at different polarization potentials. These circuit elements were used to interpret potential-step chronoamperometry results in inhibiting and passivating solutions, notably to distinguish between capacitive charging and Faradaic currents.

Chronoamperometry of Cu in acidic aqueous glycine solution containing the corrosion inhibitor benzotriazole (BTA) displayed a very consistent current decay behavior at all potentials, indicating that the rate of current decay was controlled by diffusion of BTA to the surface. In basic aqueous glycine solution, Cu (which undergoes passivation by a mechanism similar to that operating in weakly acidic hydrogen peroxide slurries) displayed similar chronoamperometric behavior for the first second or so at all anodic potentials. Thereafter, the current densities at active potentials settled to values around those expected from polarization curves, whereas the current densities at passive potentials continued to decline. Oxidized Cu species typically formed at 'active' potentials were found to cause significant current decay at active potentials and at passive potentials before more protective passive films form. This was established from galvanostatic experiments.

INTRODUCTION

We are developing a mechanistically-based tribo-chemical model for copper CMP that treats the bulk of the material removal as wear-enhanced corrosion. The model considers a copper surface to be protected by inhibitor or a protective film, depending upon the chemical nature of the slurry. The protective film is periodically removed, at least partially, by interaction with pad asperities and abrasive particles in the slurry. The corrosion current increases, then decays as the protective film builds up again until the next abrasive event. Information in the literature [1] suggests that the frequency of interactions of pad asperities with a given site on copper, τ, is about every ms.

Let $i(t')$ be the transient passive current density at time t' after bare copper is exposed to a given oxidizing passivating environment, and i_0 be the current density immediately after an abrasive-copper interaction (which would only be $i(t')$ if the interaction removed the entire film). If t is the time since the last abrasive-copper interaction, with t_0 defined such that $i(t'=t_0)=i_0$, then the average removal rate of copper between the two abrasive-copper interactions is:

$$\dot{V}_{CW} = \frac{M_{Cu}}{\rho n F \tau} \int_0^\tau i(t_0 + t)dt \qquad (1)$$

where M_{Cu} is the atomic mass of copper, ρ is the density of copper, n is the oxidation state of the oxidized copper, and F is Faraday's constant. In order to evaluate the integral, and hence to calculate the removal rate of copper, it is necessary to know the rate at which the current decreases with time. This is challenging, experimentally, because the time scale of interest is very short, on the order of ms.

Two different techniques were examined for studying copper passivation kinetics immediately after creating bare copper. The first was scratch-repassivation, whereby a copper rod is held at the potential of interest in an appropriate solution, and rotated against a sharp tool to scratch through any passive film that may be present. The current is monitored before, during and after scratching; the decay current provides the kinetics of repassivation. Unfortunately, this technique introduces more surface damage than would be expected during CMP, and hence the passivation current would be expected to be unrealistically high. Experimentally, we found that vibrations and instabilities caused significant noise, along with currents that tended to drift higher while scratching because of an increasing scratch size. Hence, we focused on the second technique, potential-step chronoamperometry. In this technique, the copper is initially held at a low potential, to allow cathodic reduction of any oxidized species that may be present on the surface. The potential is then stepped up to the anodic potential of interest, and the resulting current is monitored. When passivation occurs, or inhibitor is adsorbed onto the surface, the current decreases. Challenges in interpreting the response from potential-step chronoamperometry include Ohmic drops across the solution, the fact that the measured current includes capacitive charging in addition to the Faradaic current, diffusion that may change the composition at the copper-electrolyte interface, and rapidly changing currents that may cause potential oscillations. These effects can be reduced by using a microelectrode; the small currents minimize the Ohmic drops, and there is faster charging and diffusion. Furthermore, a microelectrode better simulates the scale of copper features during CMP than a macroelectrode.

EXPERIMENTAL PROCEDURE

Electrochemical experiments were conducted using a Gamry Series G 300 Potentiostat. The microelectrode used was a 34 gauge (80μm radius) copper wire (99.95% purity, MWS Wire Industries) insulated by enamel coating, with the exposed end acting as the microelectrode. A fresh copper surface was exposed by cutting off the end of the wire with a fine blade before each set of experiments. Before each experiment, the microelectrode was conditioned at a cathodic potential between -1.5 and -1V for 30s to remove any oxidized surface species or films.

Three different electrolytes were used: a pH 12 electrolyte containing 0.01M glycine (>99% from Acros Organics Co.) with pH controlled using NaOH (reagent grade from Fisher Scientific

Co.); a pH 4 solution buffered using acetic acid/sodium acetate with 0.01M glycine; and a similar solution also containing 0.01M benzotriazole (BTA) (from Aldrich Chemical Co.). All solutions were prepared using deionized water at 24°C. All experiments used 500 mL electrolyte

A three-electrode electrochemical cell, housed in a glass cylinder 130 mm high, with a 120 mm interior diameter, was used for all experiments. The microelectrode was suspended at the center of the cell. A saturated calomel reference electrode (SCE) [0.242 V vs the standard hydrogen electrode (SHE)] was placed 1 cm away from the wall of the cell. A platinum mesh counter electrode was held 1 cm from the wall of the cell diametrically opposite to the reference electrode. All potentials are reported with respect to the SCE. DC potentiodynamic polarization experiments were conducted from about -1V below the open circuit potential (E_{OC}) to about 1.5V above E_{OC}. Electrochemical impedance spectroscopy (EIS) was conducted at different DC potentials after holding copper at the particular DC potential for 60s, using an AC voltage amplitude of 10mV (rms). The frequency was varied from 300kHz to 0.1Hz, with 10 points per frequency decade. Chronoamperometry tests were conducted by stepping from one potential to another and recording the current densities over time at the new potential. The initial voltage, the time at initial voltage, and the final voltage were varied to examine the passivation kinetics.

RESULTS

Figures 1 and 2 show potentiodynamic polarization curves measured in pH 4 and 12 solutions containing 0.01 M glycine. At pH 4 there was active corrosion at all anodic potentials, whereas at pH 12 there was pronounced passivation between about 0 V and 0.8 V SCE. EIS measurements were taken at the potentials indicated by the arrows in these curves, and fitted to equivalent circuits to obtain equivalent electrical parameters for the different conditions.

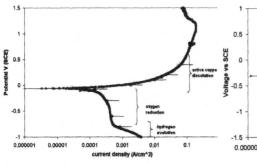

Figure 1: Potentiodynamic polarization curve (10mV/s) of a copper microelectrode in pH 4 aqueous solution with 0.01M glycine. EIS was conducted at potentials indicated by arrows

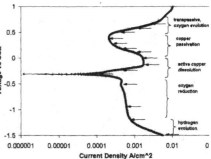

Figure 2: Potentiodynamic polarization curve (10mV/s) of a copper microelectrode in pH 12 aqueous solution with 0.01M glycine. EIS was conducted at potentials indicated by arrows

Chronoamperometry at pH 4

Knowing the equivalent electrical parameters for the different conditions allows evaluation of the capacitive charging and diffusion-limited currents in a given experiment. This then allows discernment of the Faradaic portion of the current; only this provides the passivation kinetics. To assess the quality of the electrical parameters obtained from the EIS studies, they were used to simulate potential-step chronoamperometric behavior in pH 4 aqueous glycine with no BTA, where no passivation is expected. The maximum predicted $R_U C_{DL}$ is around 3 ms, hence capacitive charging should be over by about 15 ms ($5* R_U C_{DL}$). Figure 3 shows the current predicted after stepping to 0.1V from different potentials, while Figure 4 shows experimental data for stepping to 0.2V; both final potentials were anodic, as seen in Figure 1. It is seen that by about 10 – 15 ms, the currents have settled to a steady, almost constant value; along with the good agreement between the predicted and experimental data, this demonstrates the reliability of the fitted electrical parameters.

Figure 5 shows the experimental current decay after stepping up to different potentials from -1.2V (SCE) in pH 4 aqueous glycine containing BTA. The decay behavior is very similar for all potentials. The capacitive charging is predicted to be over in less than a millisecond (EIS data show the maximum $R_U C_{DL}$ to be 0.3ms). For potentials from -0.2V to 0.4V the current decays steadily at 0.5 orders of magnitude per time decade from about 1 ms, i.e. the current density varies as the inverse of the square root of time. This suggests a Cottrell type decay behavior i.e. the current densities are diffusion limited. The behavior is very different from that in the absence of BTA[2], suggesting that diffusion of BTA to the surface is controlling the rate of inhibition. At some of the higher potentials (>0.4V), current densities stop dropping after 1s, probably because a monolayer of BTA has formed. At lower potentials (< -0.2V), the anodic currents continue to decline over time, suggesting that thick layers of BTA are forming. At these lower potentials, the cathodic reaction (hydrogen evolution) eventually becomes larger in magnitude than the diminishing anodic reaction, causing the overall current to become cathodic.

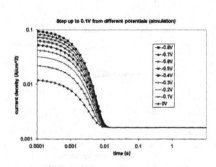
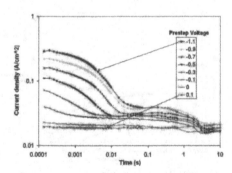

Figure 3: Simulation of current decay after stepping up from different initial potentials to fixed final potential (0.1V) for a Randles cell with electrical parameters relevant to pH 4	Figure 4: Current decay after stepping from different potentials (held for 10s) to 0.2V (fixed range data acquisition), copper in pH 4 aqueous solution containing 0.01M glycine

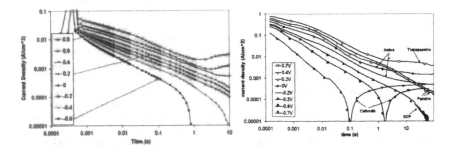

Figure 5: Current decay on copper after stepping potential from -1.2V to different potentials, copper in pH 4 aqueous solution containing 0.01M glycine and 0.01M BTA

Figure 6: Current decay after stepping from -1V to different potentials (composite data from different data acquisition modes, with data smoothing filters applied): copper in pH 12 aqueous solution containing 0.01M glycine

Chronoamperometry at pH 12

Figure 6 shows the current decay at pH 12 after stepping from -1.0V to potentials in the active, passive, transpassive and cathodic regions (although pH 12 slurries are not used in CMP, the surface pH is high when H_2O_2 is used as an oxidizing agent, so these conditions represent weakly acidic slurries containing H_2O_2). For the first second or so, the current decay at all potentials above E_{OC} is very similar, dropping by about 2 orders of magnitude from the peak current densities. After about 1 s the current densities for active or transpassive potentials settled close to the values expected from Figure 2, whereas for the passive potentials the current densities continued to drop steadily. The maximum $R_U C_{DL}$ value expected from EIS data is 0.8ms, so capacitive charging should be completely over by about 4ms (= $5*R_U C_{DL}$). Thereafter, the current should have settled to the final Faradaic value. Figure 7 confirms this; the current passing at 0.1 ms is linearly dependent on the potential step, demonstrating that it is capacitive

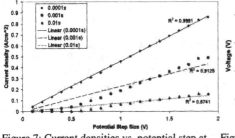

Figure 7: Current densities vs. potential step at different times after the potential step (step up from -1V to different potentials): copper in pH 12 aqueous solution containing 0.01M glycine

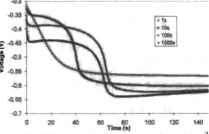

Figure 8: Galvanostatic reduction at -0.1mA/cm^2 after polarizing at a passive potential (0.2V SCE) for different times (see legend), copper working electrode in pH 12 aqueous solution containing 0.01M glycine and 10^{-4}M $CuNO_3$.

charging. At 1ms and 10ms, the relationship is no longer linear, showing that Faradaic current predominates.

The similarity in current decay between 1 ms and 1 s is interesting, in that it is seen at potentials at which copper dissolves actively and potentials where copper is passivated. Hence this behavior cannot be due to protective oxide films. Figures 8 shows galvanostatic reduction curves obtained at pH 12 on a macroelectrode that had been held at a passive potential (0.2V) for different time periods, before passing a constant reduction current of -0.1 mA/cm^2. After passivation for 10, 100 or 1000 s, the potentials dropped initially, then stayed nearly constant while the passive layers were reduced. The potential then dropped further, and hydrogen was discharged. The reduction potential of the layers decreased with increasing formation time, suggesting that the oxide stabilizes with time, requiring a higher overpotential for reduction. The layer formed over 10 s was fully reduced within 40 s, while those formed over 100 and 1000 s required longer, indicating that they were thicker. In contrast, the potential of the sample that was held at 0.2V for only 1 s dropped steadily, suggesting that there was no passive layer (instead, reduction probably involved reduction of Cu(II) ions being transported back to the electrode from the solution). This is consistent with the conclusion above that protective oxide films do not form in the first second. Instead, the decrease in current must be due to adsorption of oxidized species onto the copper. This conclusion is extremely significant for copper CMP modeling. Given the short time interval between interactions of pad asperities or abrasive particles with a given site on the copper surface (typically around 1 ms), there is not time for coherent oxide layers to develop. Hence copper CMP is best considered to be plucking of oxidized species from a film of adsorbed species, rather than as mechanical abrasion of a coherent oxidized layer.

CONCLUSIONS

Potential step chronoamperometry was used to measure the passivation kinetics of copper over very short time periods. EIS data were successfully applied to distinguish capacitive charging from the Faradaic currents relevant to material removal in CMP. Although the behavior differed in aqueous glycine solutions at pH 4 with BTA and at pH 12, in both cases the Faradaic current decreased at a well defined rate that could be incorporated into a CMP model.

ACKNOWLEDGMENTS

This work was funded by AMD, Applied Materials, ASML, Cadence, Canon, Ebara, Hitachi, IBM, Intel, KLA-Tencor, Magma, Marvell, Mentor Graphics, Novellus, Panoramic, SanDisk, Spansion, Synopsys, Tokyo Electron Limited, and Xilinx, with donations from Photronics, Toppan, and matching support by the U.C. Discovery Program.

REFERENCES

1. C.L. Elmufdi, G.P. Muldowney, "A Novel Optical Technique to Measure Pad-Wafer Contact Area in Chemical Mechanical Planarization" Mater. Res. Soc. Symp. Proc. V91, 2006 Spring
2. S. Tripathi, "Tribochemical Mechanisms of Copper Chemical Mechanical Planarization (CMP) – Fundamental Investigations and Integrated Modeling", Ph.D. Dissertation, University of California, Berkeley, December 2008.

Mater. Res. Soc. Symp. Proc. Vol. 1157 © 2009 Materials Research Society   1157-E06-03

An Investigation of the Influence of Orientation on CMP Through Nanoscratch Testing

Sarah Neyer, Burak Ozdoganlar and C. Fred Higgs III
Department of Mechanical Engineering, Carnegie Mellon University, Pittsburgh, PA 15217

ABSTRACT

With the increase in integrated circuit (IC) feature density, the surface quality obtained from chemical mechanical polishing (CMP) becomes more important as the copper interconnects decrease in size. The optimization of the IC manufacturing process will be greatly enhanced if the nanoscale effects on CMP are better understood. CMP-induced material removal (wear) at the sub-micron scale, where a single particle affects the microstructure of individual copper features within the substrate, needs to be investigated to account for wafer-scale variations. Hardness is known to affect the material removal rate, but the grain level mechanism of the removal process is not yet well known. In this work, the dependence of material removal on the crystallographic orientation of copper has been investigated by performing nanoscale scratch tests on single-crystal copper along different crystallographic directions. The surface normal of the copper crystal was indentified using orientation imaging microscopy (OIM). An analysis of the surface forces and post-scratch topography produced during the scratch tests was conducted, and the results are interpreted from a CMP perspective. Ultimately, these results are expected to refine existing material removal rate models, which do not consider the sensitivity of microstructure on the CMP process.

INTRODUCTION

Chemical Mechanical Polishing (CMP) has become the predominant planarization technique for fabricating Integrated Circuits (IC) because of its ability to create a smooth interface between the layered levels of metal networks in the IC fabrication process. While this planarization technique leads to a roughness on the order of one to tens of nanometers in copper, an understanding of the physical mechanism of planarization, or material removal, is still incomplete, especially as it relates to the microstructure (crystallography) of the metal thin-films.

The CMP process involves an IC wafer being rotated and pressed into a rough polymeric pad with slurry, made up of chemically reactive fluids and hard ceramic nanoparticles, entrained into the interface. A combination of the chemical action and the mechanical action of the nanoparticles give the polishing technique its namesake. Various mechanisms of material removal have been proposed to explain the interaction between the chemicals, the nanoparticles, the pad surface and the wafer surface and they typically fall into two categories: a primarily chemically based form of material removal which is aided by mechanical action, and a primarily chemically based form which is aided by the chemicals. The distinction between the two is that the primarily chemical models claim that the chemicals act to pull individual molecules away from the surface which is helped along by the energy of impinging nanoparticles. The primarily mechanically based models claim that the chemicals create a modified surface and nanoparticles come in and abrade it away. This paper falls into the primarily mechanical model of material removal where the nanoparticle abrades away material when a nanoparticle gets trapped between a pad asperity and a wafer asperity moving at different velocities and therefore acts to plastically deform material which ultimately becomes wear debris. Abrasion was found to be a good assumption based on calculations done by Larsen-Basse[1]. The total material removed is then the cumulative wear from all of the trapped active particles. It is assumed currently that the amount of material removed by one abrading particle is equal to the volume of the trough below the original plane of the surface [2], shown in a schematic in Fig. 1.

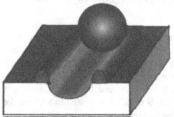

Figure 1 Schematic of a nanoparticle abrading a bulk substrate

Material removal through nanoparticle-induced abrasion is a complex and not fully-understood process. However, a thorough understanding of this process is necessary to predict the behavior of, and eventually to improve the outcome of, the CMP process. Many assumptions must be challenged when looking at size scales that approach the order of the atoms, and at forces that approach the order of molecular bonds. This paper looks at whether the material behavior at that scale can still be considered isotropic or if the bulk material behavior gives way to local behavior that can be characterized through the crystallographic arrangement of atoms.

Flom and Komanduri [3] have observed anisotropy in plastic deformation during macroscale abrasion testing with differences in material pile-up traversing from the side of the scratch to the leading edge of the scratch. In a macroscale abrasion experiment by Bailey and Gwathmey[4], the coefficient of friction and the scratch width were found to be different when scratching was performed on a copper sample in different crystallographic directions on different planes. In has also been shown that the pile-up behavior in nanoindentation depends on the crystallographic orientation of the material [5][6]. This lends itself to the belief that similar anisotropy would be found in nanoscale abrasion.

It has also been noted that the material removed during CMP changes with the crystallographic orientation of the surface in CMP of sapphire [7] and although the crystal structure and material properties are different in sapphire it is a good indicator that similar anisotropy would be found in copper CMP. The purpose of this study is to identify and characterize the anisotropy due to abrasion of nanoparticles during copper CMP using similar forces and sizes of particles found in CMP. This study aims to give a better understanding of abrasion during CMP by incorporating characteristics of local material behavior due to the crystalline structure of the material.

EXPERIMENTATION
In order to understand nanoparticle abrasion, an experiment was conducted to mimic a particle using a Hysitron Triboindenter, which is a nanoindenter that has the capability to nanoscratch as well as indent. A single scratch of Cu films at constant normal force, in similar ranges to CMP forces, would be a type of single-particle CMP. Additionally, to isolate the behavior of an individual grain within an interconnect, single crystal copper was chosen as a representative

grain. The copper specimen was nanoscratched in different sliding directions on the surface so as to investigate anisotropy in material behavior. A schematic of the experiment is shown in Fig. 2a. The behavior recorded in order to capture the resistance to material removal was the lateral force and the width and depth of the scratch was recorded to look at material removal behavior.

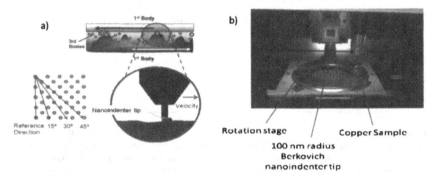

Figure 2 a) Schematic of a particle getting trapped between the pad and the wafer, which can be represented by a nanoindenter scratching in the image below. On the left is a sample surface with the four scratch test directions. b) The nanoindenter setup with rotation stage, Berkovich tip and copper sample.

In the experiment a 1 inch diameter, 1mm thick, single crystal copper specimen created by the Bridgeman technique was mounted in a cylindrical polymer base, and was polished using traditional polishing techniques in an automatic polisher. This ensured both the characterization of the orientation of the surface in an Orientation Imaging Microscope (OIM) using EBSD, and accurate material behavior during nanoscratch testing. The sample's surface orientation was then characterized in the OIM. The inverse pole figure results are in Fig. 3.

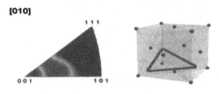

Surface Plane

Figure 3 Basic triangle with the inverse pole figure showing the sample surface plane.

In order for this set-up to work, the sample had to be placed on a rotation stage, which was firmly mounted to the Triboindenter's x-y motion stage. The nanoindenter tip used for scratching was a 100nm radius Berkovich tip which is similar in order to nanoparticle sizes in CMP. While there might be effects of the Berkovich tip shape above the point where the shape of the tip changes from a spherical into the nominal shape of the Berkovich configuration, a consistent orientation of tip the with respect to sliding direction was maintained by rotating the sample alone. The nanoindenter set-up is pictured in Fig. 2b.

EXPERIMENTAL PROCEDURE
The sample was scratched in four specified directions, each for a distance of 10μm at a constant rate of 0.33μm/s using six different forces. The inputs are listed in Table 1.

Table 1. List of input parameters to the experiment.

Normal Forces Applied					
50μN	100μN	150μN	200μN	250μN	300μN
Directions of Sliding From Reference Direction					
	0°	15°	30°	45°	
Scratch Length			Scratch Seed		
10μm			0.33μm/s		

The inputs used were found from literature on CMP nanoscratching [8]. After the initial six scratches the specimen was rotated 15° and another round of scratches was made. The sample was rotated three times. A total of 24 scratches were made on the sample. The real time lateral force was measured by the nanoindenter and the real time normal displacement was measured. The topography of each scratch was then measured by rastering the surface with the nanoindenter tip, this method was also used by Tayebi et al. [9] for similar measurements.

RESULTS AND DISCUSSION
The raw data was analyzed to determine whether anisotropy was present with respect to crystallographic orientation. An average lateral force over the 10μm scratch for each scratch was calculated. The results of each average lateral force measurement are plotted against the sliding direction for various normal forces in Fig. 4.

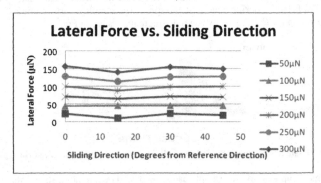

Figure 4. Plot of the average lateral force in each sliding direction for different normal forces.

Each point represents an averaged scratch and each line corresponds to the normal force that was prescribed during that scratch. It is clear that a trend is observed in the lateral force data. In the reference direction (i.e., 0°), each force is higher than the scratch of the same force 15° degrees from the reference direction. The same cycle continues for the next two 15° increments. The

88

maximum difference in lateral force between two scratches was around 20 µN which. What this corresponds to in CMP is the resistance to material removal. In a similar study by Bailey and Gwathmey [4], a cyclical trend in the data was observed in scratch tests done with forces and lengths at the micro- to macro-scales.

The average normal displacement of each nanoscratch was also averaged and plotted against sliding direction, as can be seen in Fig. 5, but it did not adhere to any cyclical trend in this data set.

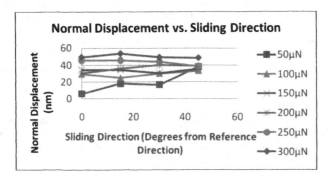

Figure 5. Plot of the average normal displacement in each sliding direction for different normal forces.

The scratch width was found by plotting the rastered surface and a manual image analysis of the graphed points was undertaken to extract the average widths of the scratch (Fig. 6).

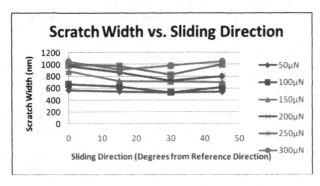

Figure 6. Plot of the average scratch width in each sliding direction for different normal forces.

The scratch width was taken to be the width across the scratch along the original surface, which does not including any extra width from the peak to peak of the pile-up. A cyclical trend was found in the scratch width data, except for a few exceptions in the data that future testing can help determine. This trend corresponds qualitatively to the macroscale scratch testing in Bailey and Gwathmey [4]. Additionally, the width of the scratch appears to be inversely correlated to the lateral force which would be anticipated, (e.g. as the scratch hardness goes up the scratch width goes down).

CONCLUSION
The representative value of resistance to material removal, the lateral force in nanoscratching, was found to be related to the crystallographic direction on the surface of the specimen. The scratch width and scratch depth were also found to vary with respect to the crystallographic direction, but more experiments are needed to determine a trend that can be modeled. Preliminary investigations of the mechanism of material removal by a nanoparticle have shown indicators that there is an anisotropic effect due the crystalline nature of materials. More work is needed to fully understand the complex nature of nanoparticle abrasion in CMP, but this single particle study suggests that microstructure could significantly impact the CMP material removal rate performance.

ACKNOWLEDGEMENTS
The authors would like to thank Dr. Anthony Rollet and Dr. Fred Lanni for their support and discussion. The research was supported by the NSF Materials Research Science and Engineering Center (MRSEC) and the Mechanical Engineering Department at Carnegie Mellon University.

REFERENCES
[1] J. Larsen-Basse, "Role of microstructure and mechanical properties in abrasion," *Scripta metallurgica*, vol. 24, 1990, pp. 821-826.
[2] Y. Zhao and L. Chang, "A micro-contact and wear model for chemical–mechanical polishing of silicon wafers," *Wear*, vol. 252, 2002, pp. 220-226.
[3] D.G. Flom and R. Komanduri, "Some indentation and sliding experiments on single crystal and polycrystalline materials," *Wear*, vol. 252, 2002, pp. 401-429.
[4] J.M. Bailey and A.T. Gwathmey, "Friction and Surface Deformation During Sliding on a Single Crystal of Copper," *Tribology Transactions*, vol. 5, 1962, pp. 45-56.
[5] Y. Liu, S. Varghese, J. Ma, M. Yoshino, H. Lu, and R. Komanduri, "Orientation effects in nanoindentation of single crystal copper," *International Journal of Plasticity*, 2008.
[6] Y. Wang, D. Raabe, C. Klüber, and F. Roters, "Orientation dependence of nanoindentation pile-up patterns and of nanoindentation microtextures in copper single crystals," *Acta Materialia*, vol. 52, 2004, pp. 2229-2238.
[7] H. Zhu, L.A. Tessaroto, R. Sabia, V.A. Greenhut, M. Smith, and D.E. Niesz, "Chemical mechanical polishing (CMP) anisotropy in sapphire," *Applied Surface Science*, vol. 236, 2004, pp. 120-130.
[8] N. Saka, T. Eusner, and J.H. Chun, "Nano-scale scratching in chemical–mechanical polishing," *CIRP Annals-Manufacturing Technology*, vol. 57, 2008, pp. 341-344.
[9] N. Tayebi, T.F. Conry, and A.A. Polycarpou, "Determination of hardness from nanoscratch experiments: Corrections for interfacial shear stress and elastic recovery," *J. Mater. Res*, vol. 18, 2003.

Mater. Res. Soc. Symp. Proc. Vol. 1157 © 2009 Materials Research Society 1157-E06-04

Novel Ceria-Polymer Composites for Reduced Defects During Oxide CMP

Cecil A. Coutinho[1], Subrahmanya R. Mudhivarthi[2,3], Ashok Kumar[2,3], and Vinay K. Gupta[1]
[1]Department of Chemical and Biomedical Engineering
[2]Nanomaterials and Nanomanufacturing Research Center
[3]Department of Mechanical Engineering
University of South Florida, Tampa, FL 33620

ABSTRACT

To meet the stringent requirements of device integration and manufacture, surface defects and mechanical stresses that arise during chemical mechanic planarization (CMP) must be reduced. Towards this end, we have synthesized multiple hybrid and composite particles on micron length scales consisting of siloxane co-polymers functionalized with inorganic nanoparticles. These particles can be easily tailored during synthesis, leading to softer or harder abrasion when desired. Upon using these particles for the planarization of silicon oxide wafers, we obtain smooth surfaces with reduced scratches and minimal particle deposition, which is an improvement from conventional abrasive materials like pure silica, ceria and alumina nanoparticle slurries. Tribological characteristics during polishing were examined using a bench top CMP tester to evaluate the in situ co-efficient of friction. Characterization of the hybrid and composite particles has been done using infrared spectroscopy, dynamic light scattering, and electron microscopy. Surface roughness of the wafers was examined using atomic force and optical microscopy while removal rate measurements were conducted using ellipsometry at multiple angles.

INTRODUCTION

Chemical mechanical polishing (CMP) is quickly becoming one of the most critical processing steps due to the reduction in feature sizes and increasing layers of metallization[1, 2]. With the advance of the semiconductor industry into the 45 nm technology node, achieving planar wafers with fewer defects after CMP is essential. Besides global planarization and high polish rate, the CMP process needs to also achieve high material selectivity and a superior surface finish. The advantages of using CMP as a global planarization technique can be nullified due to contamination from slurry chemicals, scratches, pattern related defects (dishing/erosion), delamination and dielectric crushing due to mechanical damage[3, 4]. Thus, making improvements in the CMP process to reduce the surface defects is an important engineering challenge.

Towards this end, we developed novel inorganic-organic composite microparticles that can be used in slurries and lead to improved surface characteristics after CMP. This was done by using microspherical hybrid polymeric networks, consisting of poly(N-Isopropylacrylamide) (PNIPAM) microgels with siloxane functional groups. Within these hybrid microgels are embedded nanoparticles of ceria. The presence of ceria nanoparticles has proven to be highly beneficial for oxide CMP, both in terms of removal rate and selectivity for silicon oxide[5, 6]. Incorporating the ceria nanoparticles within a polymer network reduces the friction at the polishing interface between the abrasive microparticles and the wafer surface, resulting in much fewer scratches and negligible particle embedment on the polished wafer. The combination of inorganic (e.g., siloxane and ceria) components with polymer latex particles represents a promising alternative for the synthesis of abrasive particles necessary for highly planar wafer surfaces.

EXPERIMENTAL SECTION

The composite microparticles were used to polish thermally grown oxide wafers, using a CETR CP-4 CMP bench top tester. The testing of the slurry was carried out at a 7 psi downward pressure with a 200 rpm rotation of the IC1000K groove polishing pad and a slider velocity and stroke of 3mm/s and 7mm respectively. The polishing pad was conditioned using a commercially available 3M diamond disk conditioner with a 400 grit size. The slurry flowrate was maintained at 75ml/min using a standard analog peristaltic pump.

Characterization:

The synthesis of polymer-siloxane-ceria (PSC) microcomposite particles was described elsewhere[7, 8]. The particle were examined using TEM to visually determine the extent of CeO_2 loading and dispersion within the polymer matrix. A drop of the sample solution was placed on a Formvar-coated Cu TEM grid that was examined using a FEI Morgagni 268D. The inorganic-organic composition of the composite microparticles was determined using a TA SDT Q600 thermal gravimetric analyzer (TGA). Samples were heated in air heated at a rate of 2°C/min from room temperature to 500°C.

RESULTS AND DISCUSSION

To create a robust particle, hybrid polymer-siloxane microparticles were first synthesized that contained ~10wt% silica (by TGA). As detailed in a prior publication[9], although the surfaces polished with these hybrid particles were smooth with no visible particle contamination or scratches, the oxide removal rate was low (~10-15nm/min) and needed improvement. Toward this end, we focused on the incorporation of ceria nanoparticles within the microgels. While nanoparticles of ceria are well known for their selectivity and removal of oxide from a wafer surface, they can also produce major and minor scratches that hamper further metallization[10, 11]. Therefore, microcomposites of ceria nanoparticles and microgels are attractive route towards significant improvements in the surface finish while achieving practical rates for oxide removal.

Figure 1: TEM of microcomposite

To promote incorporation of ceria nanoparticles in the microgels, we used a strategy that was successfully used in the past to form microcomposites of titania nanoparticles and PNIPAM based microgels[12]. Interpenetrating chains (IP) of poly(acrylic acid) within the hybrid microgel lead to significant fractions of carboxylic acid moieties in the microgel without altering the temperature responsive behavior and also facilitate the incorporation of ceria nanoparticles within the microgels[12]. By simply controlling the mixing ratios of the IP-hybrid microgel and ceria solutions, the mass fraction of ceria within the polymer-ceria composite particles can be easily tailored. Here the composite particles that were prepared contained approximately 50wt% ceria, which was confirmed using TGA.

The TEM image of the PNIPAM-ceria microcomposite particles (Figure 1) shows dark spots corresponding to the ceria nanoparticles (~20nm). It is evident that the PSC microcomposite particle is heavily loaded with ceria that is well dispersed and largely unaggregated within the microgels. Also, high mass loading of ceria into the IP-hybrid microgel composite, prevents free-floating ceria in the surrounding solution as shown in the above TEM image. These microcomposites were used for the planarization of 2" thermal oxide wafer. Polishing using a suspension containing 0.5wt% of the composite was compared to polishing by suspensions that contained 0.5wt% of only ceria nanoparticles. Polishing with only ceria resulted in major scratches on the wafer surface. In contrast, polishing with the composites showed no scratches and removal rates (~100nm/min) were comparable to the rates obtained with slurries containing only ceria. The lack of scratches and particle deposition on the silicon oxide wafer surface can be attributed to the cushioning effect of the polymer surrounding the inorganic oxides. The polymeric network, plausibly, helps to reduce the friction at the wafer-slurry interface and thereby, reduces surface scratches. The presence of the polymer chains can also reduce particle agglomeration that adversely affects CMP performance. Lower surface defects during polishing are useful in eliminating rigorous post CMP cleaning stages and in enabling environmentally benign CMP processes[13-15].

Quantitative thickness measurements of the oxide film on the wafer were performed using ellipsometry at multiple angles (Table 1). The PSC microcomposite particles yielded removal rates of approximately 100nm/min that was nearly 10 times higher that past studies where only hybrid microgels with no ceria were used[9]. This increase in removal rate makes it feasible to use the microcomposite particles for polishing in the final stages of CMP process where only moderate amounts of material needs to be removed but superior surface quality is required.

Table 1: Removal rates and co-efficient of friction

Slurry	COF	Removal Rate(nm/min)
0.5wt% Ceria-Polymer Composites	0.155	98
0.5wt% CeO$_2$	0.215	236
0.25wt% CeO$_2$	0.108	111

Table 1 also lists the coefficient of friction (COF) data measured during polishing. The COF was obtained from the ratio of lateral and normal forces measured in-situ using a dual force sensor installed to the upper carriage of the machine carrying the wafer carrier. The average coefficient of friction after the process had reached the steady state was been used. The average values of COF in Table 1 reveal that the slurry containing 0.25wt% ceria particles resulted in lower coefficient of friction as compared to the slurry containing 0.5wt% of ceria particles, which is presumably due to lower particle concentration. Interestingly, the composite particles lead to reduced friction at the polishing interface even though a 0.5wt% slurry of these particles should contain a higher particle concentration given the lower mass density of the organic polymer. Thus, the lower COF supports the expectation that the composite particles should have a milder abrasive interaction with the surface.

The results presented above clearly indicate that the composite particles with controlled softness/hardness can be beneficial and can be successfully implemented for polishing in the final stage of CMP process where only moderate amounts of material needs to be removed but superior surface quality is required. Fewer surface defects and particle residue[7] will aid in the elimination of rigorous post CMP cleaning stages and consequently, will help in achieving environmentally benign CMP process.

CONCLUSIONS

Composites consisting of a polymer modified with inorganic components and nanoparticles of ceria were successfully synthesized and used for low defect oxide CMP slurry applications. These particles produced a superior surface quality after polishing with very few surface scratches and no particle residue on the oxide wafer surface thereby making these particles potential candidates for next generation stringent polishing requirements.

ACKNOWLEDGEMENT

Financial support in the form of a graduate teaching assistantship from a NSF grant on Curriculum Reform (EEC-0530444) to CAC is gratefully acknowledged. The authors would also like to thank Jonathon Mbah for help with TGA analysis. Support from the University of South Florida is also acknowledged.

REFERENCES

1. M. R. Oliver, *Chemical Mechanical Planarization of Semiconductor Materials*. (2004).
2. J. M. Steigerwald, S. P. Murarka and R. J. Gutmann, *Chemical Mechanical Planarization of Microelectronic Materials*. (1996).
3. T. Y. Teo, W. L. Goh, V. S. K. Lim, L. S. Leong, T. Y. Tse and L. Chan, Journal of Vacuum Science & Technology 22 (1), 65-69 (2004).
4. L. Zhang, S. Raghavan and M. Weling, Journal of Vacuum Science & Technology 17 (5), 2248-2255 (1999).
5. D. R. Evans, Materials Research Society Symposium Proceedings 816, 245-256 (2004).
6. N. Koyama, T. Ashisawa and M. Yoshida, Patent No. JP Patent 2000109814 (2000).
7. C. A. Coutinho, R. K. Harrinauth and V. K. Gupta, Colloids and Surfaces, A: Physicochemical and Engineering Aspects 318 (1-3), 111-121 (2008).
8. V. K. Gupta, R. Jain, A. Mittal, M. Mathur and S. Sikarwar, Journal of colloid and interface science 309 (2), 464-469 (2007).
9. R. M. Subrahmanya, C. Cecil, K. Ashok and G. Vinay, ECS Transactions 3 (41), 9-19 (2007).
10. S. A. Lee, K. H. Choo, C. H. Lee, H. I. Lee, T. Hyeon, W. Choi and H. H. Kwon, Industrial & Engineering Chemistry Research 40 (7), 1712-1719 (2001).
11. Y. Tateyama, T. Hirano, T. Ono, N. Miyashita and T. Yoda, Proceedings - Electrochemical Society 26, 297-305 (2001).

12. C. A. Coutinho and V. K. Gupta, Journal of Colloid and Interface Science 315 (1), 116-122 (2007).
13. J. J. Chen, Patent No. TW Patent: 9988115709 (2001).
14. H.-W. Chiou and L.-J. Chen, IEEE International Interconnect Technology Conference Proceedings, 199-201 (1998).
15. S.-m. Jang, Y.-h. Chen and C.-h. Yu, Application:US Patent: 97810390 (1997).

Poster Session

Mater. Res. Soc. Symp. Proc. Vol. 1157 © 2009 Materials Research Society   1157-E08-06

THE EFFECTS OF HARDNESS VARIATION ON CHEMICAL MECHANICAL POLISHING OF COPPER THIN FILMS

J. Bonivel [1,2], Y. Williams[1], S. Blitz [2], A. Kumar [1]

[1] University of South Florida 4202 E. Fowler Ave., Tampa, FL 33620
[2] Carnegie Mellon University 5000 Forbes Ave., Pittsburgh, PA 15213

ABSTRACT

With the rapid change of materials systems and decreased feature size, thin film microstructure and mechanical properties have become critical parameters for microelectronics reliability. An example of a major driver of this new technology is the data storage community who is pushing for 1 Terabit/square inch on its magnetic disk hard drives. This requires inherent knowledge of the mechanical properties of materials and in depth understanding of the tribological phenomena involved in the manufacturing process. Chemical mechanical polishing (CMP) is a semi-conductor manufacturing process used to remove or planarize ultra-thin metallic, dielectric, or barrier films (copper) on silicon wafers. The material removal rate (MRR), which ultimately effects the surface topography, corresponding to CMP is given by the standard Preston Equation, which contains the load applied, the velocity of the pad, the Preston coefficient which includes chemical dependencies, and the hardness of the material. Typically the hardness, a bulk material constant, is taken as a constant throughout the wafer and thereby included in the Preston coefficient. Through metallurgy studies, on the micro and nano scale, it has been proven that the hardness is dependent upon grain size and orientation. This research served to first relate the crystallographic orientation to a specific hardness value and secondly use the hardness variation in the previously developed particle augmented mixed-lubrication (PAML) model to simulate the surface topography and MRR during CMP. Recent test and results show that currently there is no empirical formula to relate the crystallographic orientation and thereby a critically resolved shear stress (CRSS) to a specific hardness value. The second part of this investigation utilized the variation in hardness values from the initial study and incorporated these results into the PAML numerical model that incorporates all the physics of chemical mechanical polishing (CMP). Incorporation of the variation of hardness resulted in a surface topography with a difference in roughness (Ra) from the bulk constant hardness value of 60 nm. The material removal rate (MRR) of the process differs by 2.17 μm^3/s.

1. INTRODUCTION

Wear is the phenomenon of material removal from a surface due to interaction with a mating surface, either through micro fracture, chemical dissolution, or melting the contacting surface. In the case of plastic contact between hard and sharp material and a relatively softer material the hard material penetrates the softer one causing fracture; this fracture can lead to micro-cutting and ultimately material removal.

Chemical mechanical polishing (CMP) is a semi-conductor manufacturing process used to remove or planarize ultra-thin metallic, dielectric, or barrier films (layers) on silicon wafers. The material removal rate (MRR) affects the surface topography and thereby chip performance and reliability. The MRR corresponding to CMP is given by the rudimentary Preston Equation, which contains the load applied, the velocity of the pad, the Preston coefficient that includes chemical dependencies, and the hardness of the material.

For most reliability and performance tests, knowledge of the thin film constitutive mechanical behavior is required. Mechanical properties of thin films often differ from those of the bulk materials, due to the small grain sizes attributed to the deposition and various annealing methodologies. Small sized grains typically contain high grain boundary volume fractions that can lead to an increase or decrease in resulting hardness dependent on the volume fraction. This can also be partially explained by the nanocrystalline structure of thin films and the fact that these films are attached to a substrate. Most research on mechanical properties has concentrated on measurements of hardness as function of grain size, but this relationship has not been extensively investigated with in relation to CMP and the resulting MRR. Thin film mechanical properties can be measured by tensile testing of freestanding films and by the micro-beam cantilever deflection technique but the easiest way is by means of nanoindentation, since no special sample preparation is required and tests can be performed quickly and inexpensively.

1.1 MATERIAL PROPERTIES

Nanoindentation is similar to conventional hardness tests, but is performed on a much smaller scale using very sensitive load and displacement sensing equipment. The force required to press a sharp diamond indenter into tested material is recorded as a function of indentation depth. Since the depth resolution is on the order of nanometers, it is possible to indent even very thin (100 nm) films. The nanoindentation load-displacement curve provides the material's response to contact deformation. Elastic modulus and hardness are the two parameters that can be readily extracted from the nanoindentation curve. In general thin films yield stress can generally be characterized by the standard Hall-Petch relationship, due to their nanocrystalline structure, but due to the use of a metal thin film the yield stress can be taken as 1/3 or the hardness measured through nanoidentation.

1.2 CHEMICAL MECHANICAL POLISHING

In nanomanufacturing, chemical mechanical polishing (CMP) is an abrasive grinding process (AGP) used to remove or planarize ultra-thin metallic, dielectric, or barrier films (layers) on silicon wafers. During polishing a wafer is placed on a fixture and pressed against a rotating polymeric pad that is flooded with slurry. The slurry, which includes abrasive nanoparticles, polishes the films by the combined action of chemical corrosion and mechanical removal. The CMP process, shown in figure (1), is a vital interim fabrication step for integrated circuits (IC) and data storage devices where it is used to planarize thin film surfaces down to atomic smoothness to facilitate the damascene process of additive layers for lithographic patterning of wafers.

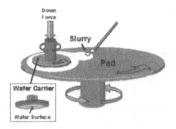

Figure 1. CMP process

An integral part of the material removal is the mechanical action of the abrasive nanoparticles and the chemical etching of the fluid phase of the slurry. Since CMP includes numerous mechanical interactions, this multi-phase tribological phenomenon is difficult to predict. Therefore, microelectronic industries have sought to advance technology by empirically "tuning" CMP to work for a material without a complete understanding of its complex aspects [1]. One major shortcoming of the existing models is that they do not account for the crystallographic effects of the materials into tribological material removal rate (MRR) relations and surface roughness predictions. Tribological MRR models mainly account for the mechanical removal [2, 3] and weakly account for the chemically-induced removal [4, 5] by using sophisticated forms of the well-known (but rudimentary) Preston Equation

$$MRR\ (x, y) = \frac{k \cdot P_{app}(x, y) \cdot U}{H} \qquad (1)$$

where k is the Preston coefficient which accounts for the chemical and mechanical removal based on polishing experiments, P_{app} is the applied pressure on the wafer, U is the relative velocity between the wafer carrier and polishing pad, H is the hardness of the material being removed (shown here as bulk constant), and MRR is the material removal rate (typically in nm/min). Hardness is the characteristic of a solid material expressing its resistance to permanent deformation. Typically the hardness is taken to be a constant throughout the wafer and then its is nondimensionalized in the Preston coefficient.

A previous model, the particle augmented mixed lubrication (PAML), model developed by the particle flow and tribology laboratory at Carnegie Mellon University, incorporates all the physics of CMP. This PAML model is dominated by the fluid mechanics and particle dynamics of the slurry, and the contact mechanics and resulting wear of the tribosurfaces [6]. Experimental validation of this model is shown in the results of the paper by Terrell et. al [6,7]. Further experimental validation will not be the crux of research here, but rather the implementation of the crystallographic hardness variation into the PAML model and a variation of the code represented in equation [2].

$$MRR\ (x, y) = \frac{k \cdot P_{app}(x, y) \cdot U}{H(x, y)} \qquad (2)$$

2.0 CRYSTALLOGRAPHY

The two major factors that effect the hardness of a material; the size of grains coupled with types of grain boundaries and the individual grain orientation, (e.g, the crystallography). The grain boundaries disrupt the movement of dislocations in a crystal and the disruption leads to larger applied forces to cause the crystal to deform. This leads to a larger yield stress for plastic deformation and thereby the smaller the grains the harder the material, this relationship is shown in the Hall-Petch equation which relates grain size to yield strength, however there is a limit to dependence on size of the grain on a micro nano-scale as the equation begins to break down for grains smaller than 1 micron and the hardness to grain size relationship on the scale has yet to be throughoughly investigated [8-13].

The orientation of the grain will determine how a dislocation will move. The presence of dislocations strongly influences many of the properties of real materials. The critically resolved shear stress (CRSS) is a characteristic property of a material and the slip system that it is activated under and can be measured by orienting a single crystal sample with respect to the applied stress and calculating the yield stress. Copper is a face centered cubic (FCC) structure which contains 12 different slip systems. The CRSS effects the yield stress of the material thereby affecting the hardness, the predominant slip plane for copper is [111] [6,8]. Values of CRSS for FCC metals range from 0.34 MPa to 0.69 MPa, for copper than value is around 0.64 MPa [14].

The orientation variation is dependent on the deposition method and more specifically the time and temperature at which the target is deposited on the subsrate. Sputtering copper directly onto silicon wafter leads to less variation while electroplating copper yields a greater variation in orientations. Annealing the wafers after the deposition process causes the variation in grain size to decrease and the predominant orientation <111> results [9].

3.0 EXPERIMENTAL DESIGN

A blanket set of polished orientation (100) 1-10 ohm-cm SSP 4850um Prime silicon wafers were deposited with .75 μm of copper using the Materials Research Science Engineering Center (MRSEC) at Carnegie Mellon University. One set was electroplated and the other set utilized sputtering as the deposition method. Both sets were then annealed at 450 degrees centigrade for 13 hours in order to allow the grains to grow on the order of several microns.

Following annealing the sample was then polished using the Strabraugh chemical mechanical polisher in order to remove any oxide layers that may have formed from the annealing process.

A Hysitron Triboidentor shown in figure [2] was then used in order to obtain the surface topography of the sputtered and electroplated surface. X-Ray diffraction and orientation imaging microscopy (OIM) were utilized after annealing to determine the orientations of the grains in the sample, these were found to be the predominant <111> orientation. A raster can was implemented in order to scan a 30-μm by 30-μm area and provide the profilomentary of that surface. Figure [3] shows the surface topography of the wafer.

Figure 2. Hysitron Nanoidentor

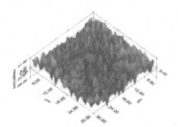

Figure 3. Surface Topography

A nanoidentation method was utilized to determine the hardness variation throughout the 30-μm by 30-μm sample and a contour plot of the hardness versus (x,y) position is shown in figure [4] along with a hardness grid corresponding to the gray scale on the plot. An indentation depth less than 10% of the film thickness was done to avoid indentation size effects (ISE) [15].

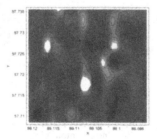

Figure 4. Contour Map of Hardness

The gray scale shown here has hardness values of 1.2 GPa, the white corresponds to values in the range of 2.0 GPa, and the black with values ranging near 1.75 GPa. Traditional average "bulk" hardness values of thin film copper are 2.5 GPa [15-17]. Elastic modulus variation was shown to range from 115 GPa to 132 GPa, which was verified by literature [15].

3.1 CMP Simulation

Several models exist that are utilized to predict wear during CMP [18-21]. These models do not incorporate a change in the microstructure property of each individual grain. The PAML model incorporates the entire multi-physics phenomenon of CMP and details of the simulation parameters are described be Terrel et.al [6]. This model is modified to incorporate microstructure change, namely the change in hardness variation per grain. The coefficients of load, P_{app}, equal to 100 micronewtons. The relative velocity is broken into the velocities of the lower pad, U_{pad} and the wafer carrier equal to 10 rpms each. The Preston coefficients, k equal to 1, are all initialized in the PAML model. The Preston coefficient is set to 1 therefore the effects of the slurry are negated for this simulation. The authors determined a thorough understanding of the mechanical abrasion and the resulting effect from variable hardness is a first case scenario. Future simulations will incorporate slurry chemistry and colloidal particle effects. The experimental surface topography was imported and a technique developed by Dickerell et. al, is utilized to convert the experimental data into volume pixels (voxels) [22]. The surface topography from figure (3) is then turned into voxels and each voxel, shown in figure (5) contains the x-y position in three-dimensional space of each grain along with the corresponding value of hardness from experiments. A random pad surface topography is generated and contact is initiated with the wafer surface. The stress on individual voxels is calculated and these stresses are used to calculate individual and cumulative wear rates on each voxel and cumulatively on the wafer surface. The wear distance from each voxel is calculated and then subtracted from the wafer surface. The resulting surface voxel height is then adjusted in each step and a resulting surface topography can be calculated at each step. The roughness (Ra) is calculated from the cumulative voxel height from the resulting surface topography and verified the experimental procedure.

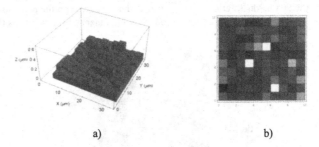

a)   b)

Figure 5. Voxel Representation of Surface Topography a) voxel surface topograph b) voxel hardness contour plot

3.2 RESULTS

The linear relationship between polishing time and wear is still evident from figure (6). As the time steps are increased, the resulting wear is increased which follows until the entire surface would be polished and thereby the wear would evidentially decrease.

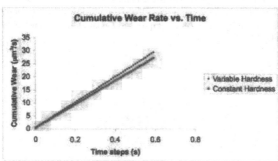

Figure 6. Cumulative Wear rate of CMP

The MRR difference between the variable hardness and bulk modulus value (constant hardness) is 2.17 $\mu m^3/s$; this rate is due strictly to mechanical abrasion, the chemical effects and particle dynamics of the slurry have not been incorporated in this initial investigation. Of greater importance is the resulting surface topography as this relates directly to the viability of the integrated circuit or media storage device fabrication. Figure [7] shows the comparison of surface topographies. Figure [7a] shows the resulting topography of the simulation run with constant bulk hardness with a value of 1.5 GPa, which is currently utilized in previous models. Figure [7b] shows the resulting topography from the model with the true hardness values taking from nanoidentation of individual grains shown in figure [4].

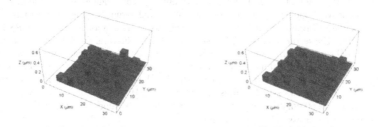

a) Constant hardness   b) Variable hardness

Figure 7. Simulation surface topography comparison

Incorporation of the variation of hardness resulted in a surface topography with a difference in the resulting roughness (Ra) values of 60 nm. This value represents the 30 ms of runtime. The constant hardness simulation had a surface roughness of 233 nm while the variable hardness simulation was 173 nm. The variable hardness simulation has a smoother surface due to the fact that many of the individual grains within the simulation were below the bulk constant of 2.5 GPa, and this resulted in more material removal for each grain below the threshold. The constant hardness removed the same amount of material for each grain that was in contact with the polishing pad surface at a given time step, this resulted in a rougher overall surface.

The relative velocity of the pad and were also varied to view the surface topographies that would result from the simulation and see if the would follow previous model parameters. Figure [8] shows the resulting surface topography for the same initial conditions but with the U_{pad} and U_{wafer} set to 20 rpm.

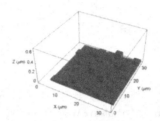

Figure 8. Simulation with variable hardness and increased velocity

The resulting topography follow in conjunction with the Preston equation [2] in reference to the higher the speed or pressure applied, the more material that is removed. The surface was polished at a much quicker rate and the resulting in surface roughness of 108 nm.

4.0 CONCLUSION

A two-part investigation was conducted in order to determine if a previously developed chemical mechanical polishing (CMP) model, PAML, could be enhanced through further simulations. The first part involved relating the critically resolved shear stress (CRSS) of a single crystal to an individual hardness value. An investigation relating the CRSS to the hardness value was conducted based on the orientations and hardness values from experimentally found properties. Currently, there is not an empirical model or equation to relate the CRSS to the hardness value. The second part of this investigation utilized the variation in hardness values from the initial study and incorporated these results into a particle augmented mixed-lubrication (PAML) numerical model that incorporates all the physics of chemical mechanical polishing (CMP). Inclusion of the hardness variation into the PAML model differentiates it from current models that only utilize the bulk hardness of the film for CMP. The surface topography simulations show a 60 nm difference in the resulting roughness values.   The material removal rate (MRR) of the process differs by 2.17 μm^3/s.

Evidence of the importance of the variation in hardness and its results on the surface topography and MRR during CMP are proven through this work. The results from the topography variation have implications of integrated circuit failure if not properly presided over, as the atomic roughness is instrumental in the fabrication of integrated circuits as well as patterned media storage devices. This work was a first step in the experimental validation of PAML code with microstructure incorporated versus the bulk material properties being taken.

The next step in this work is the incorporation of the fluid and particle dynamics in the slurry application to measure this dependency on the resulting MRR and surface roughness.

Acknowledgments
The authors would like to thank Dr. C. Fred Higgs III and Dr. Elon Terrell for their use of the PAML code and allowing for modification of the code for simulation purposes. The author would also like to thank the National Science Foundation who helped to fund this work, the Materials Research and Science Engineering Center at Carnegie Mellon University, and the Nanomanufacturing and Nanomaterials Research Laboratory at the University of South Florida.

REFERENCES

1. D.T. Read, J.W. Dally, J. Mater. Res., 8(7), p. 1542-1549, 1993

2. T.P. Weihs, S. Hong, J.C. Bravman, and W.D. Nix, J. Mater. Res. 3(5), p. 931-942, 1988

3. S.P. Baker and W.D. Nix, J. Mater. Res. 9(12), p. 3131-3144, 1994

4. S.P. Baker and W.D. Nix, J. Mater. Res. 9(12), p. 3145-3152, 1994

5. M. Doerner and W. Nix, J. Mater. Res. 1, p. 601, 1986

6. E. J. Terrell, and C. F. Higgs III, J. of Tribology, vol. 131(1), p. 012201 – 012210, 2009

7. E.J, Terrell, C.F. Higgs III, ASME Paper IJTC2007-44487, 2007

8. J. Chen, W. Wang, L.H., Qian, and K. Lu, *Scripta Materialia*, v 49, n 7, p 645-50, 2003

9. K. Barmak, C. Cabral J. Vac. Sci. Technol. B, Vol. 24, No. 6, Nov/Dec p. 2485-2498, 2006

10. M. Ignat, P. Scafidi, E.Duloisy, and J. Dijon, *Materials Reliability in Microelectronics IV. Symposium*, 135-46, 1994

11. T. Fuji and Y. Akiniwa, *Key Engineering Materials*, v 340-341, pt.2, 979-84, 2007

12. H. Conrad, J. Narayan, *Scripta Mater* 42(11):1025–30, 2000

13. J. Schiotz, F.D. Di Tolla, K.W. Jacobsen, Nature 391 p.561,1998

14. W.D. Callister. Fundamentals of Materials Science and Engineering, 2nd ed. Wiley & Sons.

15. A.A. Volinsky, J. Vella , I.S. Adhihetty , V. Sarihan , et.al, J. Mater. Res. 5(3) p. 1-6, 2001.

16. S. Suresh, T.G. Nieh, and B.W. Choi, *Scripta Materialia*, Vol. 41, No. 9, p. 951–957, 1999

17. T. Fang, and W. Chang, Microelectronic Engineering 65 p. 231–238, 2003

18. G., Nanz, L.E, Camilletti, IEEE Trans. Semicond. Manuf., 8(4), p. 382–389, 1995

19. P.B., Zantye, P. B., A., A.K. Sikder, Mater. Sci. Eng., R., 45(36), pp. 89–220, 2004.

20. D., Castillo-Mejia, and S., Beaudoin, J. Electrochem. Soc., 150(2), pp. G96–G102. 2003

21. J., Seok, C.P., Sukam, A.T., Kim, J.A., Tichy, and T.S. Cale, J. of Wear,254(34), p 307–320, 2003.

22. D.J. Dickrell, Dugger, M.T., Hamiltion, M.A., and Sawyer, W.G, J. Micromech Sys. 16(5), p. 1263-1268, 2007

23. D. Beegan, S. Chowdhury, and M.T. Laugier, Surface Coatings Tech 201, p. 5804-5808, 2007

24. Y. Liu, A.H.W. Ngan, Scripta Materialia Vol. 44, p.237-241, 2001

25. Z.W. Shan , R.K. Mishra , S.A. Syed, O.L. Warren, A.Minor, Nature Mater. 7 p.115, 2008

26. W. Tseng, C. Liu, B. Dai, C. Yeh, Thin Solid Films Vol. 290-201, p. 458-463, 1996

Mater. Res. Soc. Symp. Proc. Vol. 1157 © 2009 Materials Research Society   1157.F08-07

Novel Method to Synthesize Ceria Coated Silica Particles

Myoung-Hwan Oh*, Jae-Seok Lee, Sushant Gupta, Tae-gon Kim, Aniroddh Kaanna and Rajiv K. Singh
Department of Materials Science and Engineering, University of Florida, Gainesville, Florida, 32611, U.S.A.

ABSTRACT

Monodispersed ceria coated silica particles were prepared by a new type of ceria precursor. The ceria precursor was synthesized by alkoxide method, which employs ethanol as solvent. The synthesized particles were characterized with scanning electron microscopy (SEM), transmission electron microscopy (TEM), X-ray diffraction (XRD) and X-ray photoelectron spectroscopy (XPS). It was found that well-crystalline ceria coatings were deposited on the surface of the silica particles without post-heat treatment. In addition, the coated particles prepared by a new precursor were uniformly dispersed without the formation of hard aggregate as compared to those obtained by conventional method.

INTRODUCTION

The ceria (CeO_2) particles have been widely used as a slurry abrasive for the STI CMP process in the semiconductor industry. The ceria based slurries provide excellent CMP performance owing to high nitride/oxide selectivity, chip uniformity, global planarization and high polishing efficiency [1, 2]. However, the ceria-based slurries induce defects on the wafer surface because of the broader particle size distribution and agglomeration particles resulting from poor dispersion stability. In order to overcome these problems, many researchers have investigated methods for producing ceria coated silica particles because colloidal silica offer certain advantages such as monodispersed spheres, narrow size distribution and lower specific gravity [3, 4]. However, in previous works [5-7], the synthesized ceria coated silica particles undergo severe aggregation from large volume shrinkage during the drying process because cerium sources induce the wet precipitate gelatinous in precipitation process [8]. Moreover, the resulting dispersions contain both of the ceria coated silica particles and the hard aggregates of ceria particles. In addition, in some cases, certain particles need the post-heat temperature to acquire a pure ceria coating.

The objective of this study is to prepare the ceria coated silica particles without hard aggregates and verify the formation of ceria coating. This study is focused on the effect of ceria precursor on the formation of ceria coating. The evidences of pure ceria coated on the surface of silica particles are examined by means of XRD, SEM, TEM, and XPS.

EXPERIMENTAL DETAILS

To prepare ceria precursor, cerium (III) nitrate hexahydrate ($Ce(NO_3)_36H_2O$) and sodium hydroxide (NaOH) were used as the starting materials. Cerium (III) nitrate hexahydrate (1 g) and sodium hydroxide (1 g) were separately dissolved in 20 ml ethanol and then mixed vigorously for 12 h at 50°C. Hydrogen peroxide (H_2O_2) was added to the mixing solution as an oxidizing agent. The precipitated ceria precursor was separated via centrifugation, and then dispersed in 80 ml of water under continuous stirring. The pH of the suspension was adjusted to under the 0.1 by adding nitric acid (HNO_3). Finally, the solution was heated to 40 °C for 10 min. After reaction, a transparent light yellow solution was obtained which was then cooled to the room temperature.

The ceria coated silica particles were prepared by admixing 2 ml of ceria precursor with 0.1 g of monodispersed silica particles synthesized by the modified Stöber method [4]. Silica particles were dispersed in 30 ml of water and then ceria precursor was added into dispersed silica solution. The mixed solution was stirred for 5 h at 60 °C after adjusting the pH to 6.0 ~ 7.0 with ammonium hydroxide (NH_4OH). The precipitated particles were washed and dried in an oven.

In order to compare, the ceria coated silica particles were also prepared by conventional method [5-7]. 0.1 g of silica particles were dispersed in 30 ml of water and then 2.6 g of cerium (III) nitrate hexahydrate was added into the dispersed silica solution. The mixed solution was stirred for 6 h at 60 °C after adjusting to pH 7.0 with ammonium hydroxide.

The morphology and sizes of particles were examined by Field Emission Scanning Electric Microscope (FE-SEM). The crystalline phase identified x-ray diffraction (XRD) using CuKα radiation. The XPS measurements were performed on an ESCA with a monochromatic Mg Kα source and a charge neutralizer.

RESULTS AND DISCUSSION

XRD patterns of the bare silica particles and ceria coated silica particles synthesized from the two methods are shown in Fig. 1. The curve (a) represents the XRD spectra of the bare silica

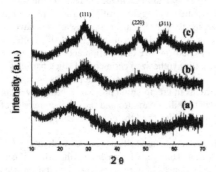

Figure 1. XRD patterns of the synthesized particles; (a) bare silica particles, (b) ceria coated silica particles prepared by conventional precipitation method and (c) using a new type of ceria precursor.

particles while curve (b) and (c) show XRD pattern of the coated silica particles with conventional precipitation and alkoxide method, respectively. The major reflections associated with fluorite structure of ceria coating can be observed from curve (b) and (c) in Fig. 1. However, the ceria coating synthesized by alkoxide method show higher peak intensity as compared to that prepared by conventional precipitation method. It is interesting to note that the ceria coating synthesized by using ceria precursor shows better crystallinity in spite of the processing temperature being same as used in the conventional precipitation method.

Fig. 2 shows the FESEM micrographs of the bare silica particles and ceria coated silica particles prepared from the two methods. Fig. 2(a) shows the smooth surface of the bare silica particles in contrast to the rough and texture surface of the ceria coated particles shown in Fig. 2(c) and (d). Moreover, as shown in Fig. 2 (b) and (d), the different in morphology between two specimens is very interesting. The ceria coated silica particles obtained from conventional

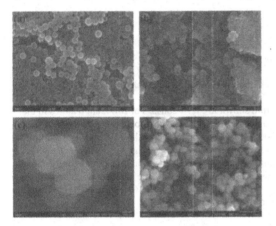

Figure 2. FESEM micrographs for (a) bare silica particles, (b) ceria coated silica particles prepared by conventional precipitation method, (c) and (d) using a new type of ceria precursor.

precipitation method were severely aggregated with a massive ceria layer. It appears that this result is mainly due to the bridging of adjacent particles with water by hydrogen bonding and the subsequent high capillary force during the drying process [9]. On the other hand, the ceria coated silica particles synthesized from ceria precursor were monodispersed without hard aggregates. This result is consistent with the study of Ikegami et al. [10], which shows that the $Al(OH)_3$ and $Y(OH)_3$ wet precipitates show better dispersity in alcohols and the dispersing effect increases with the increasing carbon chain length of the alcohols. In this works, the carbon chain of ceria precursor formed from ethanol based solvent helps to decrease the inter-particle attraction forces during the drying process and hence increase the dispersion of the ceria coated silica particles.

The XPS analysis was used to determine the oxidation state of coating deposited on the surface of silica particles. The O 1s core spectra are very informative terms of the structural role of ceria coating. As seen from the Fig. 3, it is clear that O 1s peak can be deconvoluted into two peaks associated with silica and ceria. According to the data published by National Institute of

Standards and Technology of USA, the peak at 532.4 eV is attributed to the silica and the other at 529.8 eV is associated with the ceria.

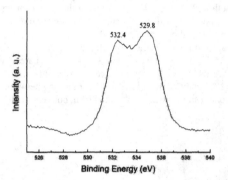

Figure 3. XPS spectra for O 1s core peak of ceria coated silica particles.

Moreover, the XPS Ce 3d spectrum shown in Fig. 4 consists of two main features in the range of 870 ~ 930 eV, corresponding to the Ce $3d_{3/2}$ and $3d_{5/2}$ components due to the spin-obit coupling. In previous works [11, 12], both Ce(III) and Ce(IV) show the $3d_{5/2}$ and $3d_{3/2}$ multiplets. However, Ce(III) shows only two peaks for each component ($3d_{5/2}$ and $3d_{3/2}$) whereas Ce(IV) shows three peaks associated with the initial state of tetravalent cerium. In addition, the presence of the initial state of tetravalent cerium is further substantiated by the evidence of an additional peak at a binding energy of 918 eV. The presence of three structures (P1, P2 and P3) confirms the coating of ceria on the surface of silica particles.

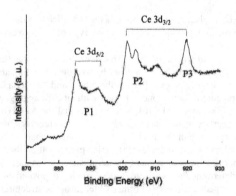

Figure 4. XPS Ce 3d multiplex of ceria coated silica particles.

CONCLUSIONS

Monodispersed ceria coated silica particles were prepared by using ceria precursor synthesized from alkoxide method. The ceria coated silica particles prepared by the new precursor showed better crystallinity without heat treatment temperature. The particles were uniformly dispersed and did not exhibit the formation of hard agglomerates. Further it was confirmed from XPS analysis that the ceria coating on silica particles composed of CeO_2 phase and no Ce_2O_3 phase was observed.

REFFERENCES

1. T Hoshino, T. Kurata, Y. Terasaki, and K. Susa, *J. Non-Cryst. Solids* **283**, 129 (2001).
2. D. S. Lim, J. W. Ahn, H. S. Park, and J. H. Shin, *Surf. Coat. Technol.* **200**, 1751 (2005).
3. A. Yoshida, "The Colloidal Chemistry of Silica," Advance in Chemistry Series, vol. 234, (Oxford University Press, Oxford, 1994) pp. 51–62.
4. W. Stöber, A. Fink, and E. Bohn, *J. Colloid Interface Sci.* **26**, 62 (1968).
5. M. S. Tsai, *Mater. Sci. Eng.: B* **104**, 63 (2003).
6. S. H. Lee, Z. Lu, S. V. Babu, and E. Matijevic, *J. Mater. Res.* **17[10]**, 2744 (2002).
7. X. Song, N. Jiang, Y. Li, D. Xu, and G. Qiu, *Mater. Chem. Phy.* **110**, 128 (2008).
8. J. G. Li, T. Ikegami, J. H. Lee, and T. Mori, *Acta Mater.* **49**, 419 (2001).
9. M. S. Kaliszewski, and A. H. Heuer, *J. Am. Ceram. Soc.* **73**, 1504 (1990).
10. T. Ikegami, and N. Sati, *J. Ceram. Soc. Jpn.* **104**, 469 (1996).
11. N. Thromat, M. Gautier, and G. Bordier, *Suf. Sci.* **345**, 290 (1996).
12. S. Gavarini, M. J. Guittet, P. Trocellier, M. Gautier-Soyer, F. Carrot, and G. Matzen, *J. Nuclear Mater.* **322**, 111 (2003).

Mater. Res. Soc. Symp. Proc. Vol. 1157 © 2009 Materials Research Society 1157-E08-08

Study of Conditioner Abrasives in Chemical Mechanical Planarization

Chhavi Manocha[1], Ashok Kumar[2], Vinay K. Gupta[1]

[1]Department of Chemical Engineering, University of South Florida, Tampa, FL 33613, U.S.A.
[2]Department of Mechanical Engineering, University of South Florida, Tampa, FL 33613, U.S.A.

ABSTRACT

Chemical Mechanical Planarization (CMP) has emerged as the central technology for polishing wafers in the semiconductor manufacturing industry to make integrated multi-level devices. Both chemical and mechanical processes work simultaneously to achieve local and global planarization. Although extensive research has been carried out to understand the various factors affecting the CMP process, many aspects remain unaddressed. One such aspect of CMP is the role of abrasives in the process of conditioning. Abrasives play an important role during conditioning to regenerate the clogged polishing pads. This research is focused on the study of abrasives in the process of conditioning with a focus on the size of abrasives. With diamond being widely used as an abrasive for conditioning the polishing pad, five different sizes of diamonds ranging from 0.25µm to 100µm were selected to condition the commercially available IC 1000 polishing pad. Properties like pad roughness and pad wear were measured to understand the effect of the abrasive size on the pad morphology and pad topography. In-situ 'coefficient of friction' was also monitored on the CETR bench top tester. The final impact was seen in the form of surface defects on the polished copper wafers using optical microscopy.

INTRODUCTION

The CMP polisher, as shown in Figure 1, consists of a bottom rotating platen on which a polishing pad is fixed. The top platen holds the wafer to be polished, face down on the pad. During the polishing run, the wafer comes in contact with the pad surface with a set downforce. Both the platens rotate during the run. Polishing slurry consisting of abrasive particles and other necessary chemical constituents are fed to the pad surface to carry out the polishing where the pad comes in contact with the wafer being polished. The abrasive slurry has several functions. The pad used for polishing is seen to glaze with time under the effect of polishing. To get higher removal rates, the pad needs to be regenerated. This is done by conditioning, which is the second stage of the CMP process. The rotating polishing pad is abraded using a conditioner which is an abrasive disk that opens up the clogged pores of the pad (*1-3*)

The polishing mechanism can be explained on the basis of the contacts occurring in the system. The chemicals in the slurry, on contact with the wafer, soften its surface, while the abrasives in the slurry remove this layer. In the absence of the mechanical fraction, the chemical effect is limited (*4*)

Figure 1: CMP bench top tester

EXPERIMENTAL SETUP

Towards the objective of evaluating the effect of abrasive size on the pad morphology(5), three parameters were varied. These parameters include abrasive size, time of conditioning and rotational speed of the polishing pad. The polishing pad used was IC 1000 and sub-pad used was SUBA IV. Deionized (DI) water was used as fluid in these conditioning experiments. Five different sizes of diamond abrasives ranging from 0.25μm to 100μm were chosen to carry out the conditioning on the commercially available polishing pad. The different sizes of diamond abrasives used were 0.25μm, 2μm, 8μm, 68μm, and 100μm. Though all these conditioners appear alike to the naked eye, as in Fig 2, the different diamond sizes can be clearly distinguished in the corresponding SEM images as seen in Figure 3. In order to evaluate the influence of abrasive size on pad morphology, conditioning was carried out for four different time durations: five, ten, fifteen, and twenty minutes. In addition, each experiment was conducted at two rotation speeds (150 rpm and 200 rpm) of the pad.

To carry out the polishing experiments for copper wafers, three different abrasive sizes for conditioners were used; 8μm, 68μm, and 100μm to condition the polishing pad. The slurry used was Cabot 5001.

(a)   (b)   (c)   (d)   (e)

Figure 2: Photographs of conditioners with different abrasive sizes
(a) 0.25μm (b) 2μm (c) 8μm (d) 68μm (e) 100μm

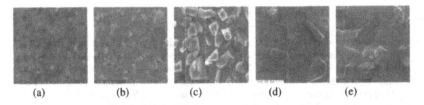

(a)   (b)   (c)   (d)   (e)

Figure 3: Photographs of conditioners with different abrasive sizes
(a) 0.25µm (b) 2µm (c) 8µm (d) 68µm (e) 100µm

DISCUSSION

Pad Wear

The reduction in thickness of the pad as the conditioning proceeds is defined as pad wear. The conditioner has to maintain constant contact with the pad throughout the conditioning process. In order to maintain constant contact, the carriage automatically comes down by an amount equal to the decrease in the thickness of the pad. The distance by which the carriage travels downward for the conditioner to maintain constant contact with the pad is a measure of the amount of wear in the pad.

Figure 4 show plots of the amount of pad wear against time for different abrasive sizes at a pad rotation speed of 150 and 200 rpm. From Figure 4(a) it is observed that for larger abrasive, the amount of wear in the pad is higher than that for smaller abrasive. For example, for an abrasive size of 100µm the amount of wear at five minutes is 0.018µm whereas for an abrasive size of 0.25µm, the amount of wear is 0.006µm. As the time for conditioning increases, the amount of wear also increases. The same trend is observed in Figure 4(b) when the pad rotation speed is set at 200 rpm. However, the amount of wear is at a rotation speed of 200 rpm higher when compared to the amount of wear at 150 rpm. At 150 rpm, the amount of wear when using 8µm abrasive size for a conditioning time of 20 minutes is 0.015µm, whereas at 200 rpm it is 0.016µm. As larger abrasives abrade the pad to a higher degree than the smaller ones, the pad wear caused due to this abrasion is also greater in the case of bigger abrasives. Higher the pad wear, shorter is the pad life. An uneven pad wear also results in uneven slurry transport and non-uniform polishing.

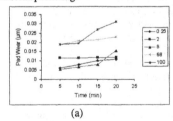
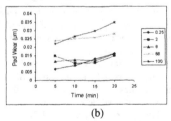

(a)   (b)

Figure 4: Plot of pad wear at different conditioning times for different abrasive sizes
(a) 150 rpm (b) 200 rpm

Coefficient of friction

The next step is to evaluate the coefficient of friction resulting from the different abrasive sizes. During the conditioning process, the lower platen exerts a normal force equivalent to the downward force experienced by the pad. There is a shear force also present between the pad and conditioner. The coefficient of friction between the pad and the conditioner is computed as a ratio of the shear force to the normal force. Figure 5(a) and Figure 5(b) show the plots of coefficient of friction against time for the conditioning runs at 150 and 200 rpm respectively. From the plots it is observed that with an increase in abrasive size, the coefficient of friction increases. As discussed, pad wear is more at higher pad rotation speeds and hence there is a decrease in shear force between the pad and the conditioner. This further implies that the coefficient of friction should decrease as the pad rotation speed increases. This can be seen in the plots below. For an abrasive size of 100μm, the average coefficient of friction at 150 rpm is 0.624 whereas at 200 rpm the average coefficient of friction is 0.589. Similarly, when the abrasive size is 2μm, the average COF at 150 rpm is 0.44 whereas at 200 rpm, the COF is 0.40.

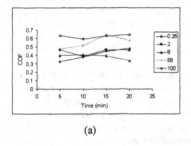 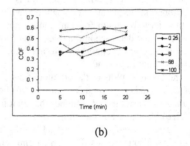

(a)   (b)

Figure 5: Plot of COF at different conditioning times for different abrasive sizes
(a) 150 rpm (b) 200 rpm

Pad roughness

The roughness of the pad affects the removal rate as the wafer being polished comes in direct contact with the rough surface of the pad. Having evaluated the coefficient of friction, the next step is to evaluate the roughness of the pad. This was done using a contact mode profilometer with a stylus of radius 12μm. A standard scan was conducted over a 50μm scan length. There were five measurements for each pad. The average roughness measurements of each pad are plotted against the abrasive size of the conditioner used for that pad. Figure 6(a) shows the plot of pad roughness with respect to time for different abrasive sizes at 150 rpm. It can be seen from the plot that the roughness increases with time. For a conditioning time of 5 minutes, the pad roughness values were 0.775, 1.03 and 1.26 μm for abrasive sizes of 0.25μm, 2μm and 8μm, respectively while for a conditioning time of 20 minutes, the roughness values for 0.25μm, 2μm and 8μm abrasives are 1.23, 1.28 and 2.05μm, respectively. Conditioning with bigger abrasives results in higher roughness compared to the smaller abrasives. As seen from the plot, for a conditioning run of 20 minutes, the pad roughness for 68μm is still increasing while for that for 100μm seems to decrease. Figure 6(b) shows the plot of pad roughness with respect to time for different sizes at 200 rpm. On comparison of roughness values obtained at 200 rpm with those at 150 rpm, it is observed that the pad roughness values at 150 rpm are higher than those at 200 rpm. This could be due to the fact that at a higher rotational speed, the pad does not get abraded uniformly.

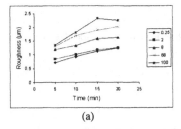 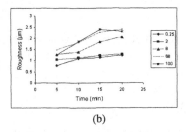

(a)   (b)

Figure 6: Plot of pad roughness at different conditioning times for different abrasive sizes
(a) 150 rpm (b) 200 rpm

Optical images of polished wafers

For the experiments, the wafer was polished using a pad, the pad was conditioned again before reusing it to polish the next wafer. The process was repeated for three different wafers. The results are shown in Figure 7 where Figure 7(a), (b) and (c) show the optical images of the first, second and third wafer polished using the pad conditioned with 8μm, 68μm, 100μm abrasive size respectively. On comparing the optical images it is observed that the pad conditioned using 100μm abrasive size conditioner produced a large number of scratches on the wafer while that conditioned with the 68μm abrasive conditioner produced fewer number of scratches. This implies that the bigger conditioner abrasive results in more scratches during polishing. It was also seen that the number of wafer scratches in case of 100μm and 68μm conditioner decreased from the first wafer to the third wafer while the number of scratches roughly remained the same in case of wafer polished on the pad conditioned with the 8μm conditioner abrasive.

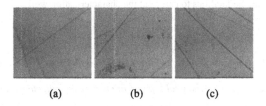

(a)   (b)   (c)

Figure7(i): Optical images of the first wafer polished using pad conditioned with different abrasive sizes.
(a) 8μm (b) 68μm (c) 100μm

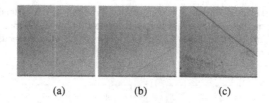

(a)   (b)   (c)

Figure 7(ii): Optical images of the third wafer polished using pad conditioned with different abrasive sizes.
(a) 8μm (b) 68μm (c) 100μm

CONCLUSION

A systematic study of conditioner abrasives ranging in size from 0.25 μm to 100 μm was performed. The results showed that the conditioning with bigger abrasives gave higher pad roughness. The roughness value was also found to increase with an increase in conditioning time. However, the larger 100 μm abrasives, when used for longer conditioning times, start to rupture the pad. Also, for these larger abrasives, longer conditioning times led to higher pad wear, which has important consequences such as reduced pad life. The 68 μm and 100 μm conditioner abrasives also showed a high value of COF (0.5-0.7) during conditioning, which was nearly twice the value found for the smaller abrasives. Though the smaller abrasives showed a steady increase in the COF values with an increase in conditioning time, the change in COF value for bigger abrasive was nearly negligible. The results of the systematic study in this work indicate that smaller abrasives need longer conditioning times to achieve same degree of pad abrasion as the bigger conditioning abrasives.

ACKNOWLEDGEMENTS

I would like to thank NSF for their support to carry out the research work. I would also like to thank Nanomaterials and Nanomanufacturing Research Centre at University of South Florida for providing the facilities to carry out the characterization work.

REFERENCES

1. B. J. Hooper, G. Byrne, S. Galligan, *Journal of Materials Processing Technology* **123**, 107 (2002).
2. M. R. Oliver, *Chemical-Mechanical Planarization of Semiconductor Materials.* (Springer, 2004).
3. J. M. Steigerwald, S. P. Murarka, R. J. Gutmann, *Chemical Mechanical Planarization of Microelectronic Materials.* (Wiley-Interscience, 1997).
4. P. B. Zantye, A. Kumar, A. K. Sikder, *Materials Science & Engineering R* **45**, 89 (2004).
5. T. Sun, L. Borucki, Y. Zhuang, A. Philipossian, *Materials Research Society Symposium Proceedings* **991**, 45 (2007).

**CMP in Memory and
Data Storage Technologies**

Mater. Res. Soc. Symp. Proc. Vol. 1157 © 2009 Materials Research Society   1157.F10-01

Issues and Challenges of Chemical Mechanical Polishing for Nano-Scale Memory Manufacturing

Choon Kun Ryu, Jonghan Shin, Hyungsoon Park, Nohjung Kwak, Kwon Hong and Sung Ki Park

Advanced Process Technology Group, R&D Division, Hynix Semiconductor Inc. Icheon-si, Kyoungki-do, Korea 467-701

ABSTRACT

As memory devices shrink down to nanoscale, the number of the CMP process has increased gradually due to the complexity of integration scheme. The CMP for isolation has increased significantly because the isolation process of metal contact plugs and damascene metallization at nanoscale has been successfully enabled by the CMP. The CMP selectivity, which depends strongly on the chemistry of the slurry, must be tuned for the various new materials. Recently, in order to get over the limitation in lateral shrinkage of the memory device, several novel vertical integration approaches such as Wafer-direct-bonding and Through-Silicon-Via have been investigated extensively. The vertical integration needs the new CMP process such as high removal rate Cu CMP. Next generation memories need the CMP process for new materials such as GeSbTe, conductive oxide, and magnetic materials. Since any nano-size scratch will be a killer defect at the nanoscale memory, both the CMP equipment and the consumables must be maintained with tighter degree of control specifications.

INTRODUCTION

As memory devices shrink down to nanoscale, there have been a variety of issues and challenges in both device performance and process integration. The CMP process also has faced lots of hurdles toward the successful nanoscale process integration. The CMP processes for global and local planarization have been implemented to replace conventional planarization processes since the planarity requirement for nano-lithography becomes more stringent. There is a great demand on the reduction of cost-of-ownership since the addition of the CMP process causes the rise of the manufacturing cost. The CMP process has evolved for a variety of new materials and integration schemes which have been introduced to resolve critical issues of device performance and patterning in nanoscale regime.[1] In this paper, current issues and challenges of the CMP in nano-scale memory manufacturing are discussed.

EVOLUTION OF CMP PROCESS IN MEMORY MANUFACTURING

As shown in figure 1, the number of the CMP process has increased gradually due to the complexity of integration scheme. The DRAM has much more CMP processes than the flash memory because the DRAM has more complex topology due to the capacitor memory cell. Notably, the number of CMP processes increased exponentially below 70 nm tech node of DRAM. The CMP process in the NAND Flash memory increased gradually from 120 nm to 50 nm and then saturated below 50 nm. When new materials or new device structures were introduced to overcome the degradation of the device performance due to the tech shrinkage, additional CMP processes have been developed also. When the metal bit line was patterned by a damascene method in order to overcome the patterning hurdle of the reactive ion etching, a metal CMP process was added. As shown in figure 2, the CMP for isolation has increased significantly with respect to the shrinkage of the tech node because the isolation process of metal contact plugs and damascene metallization at nanoscale has been successfully enabled by the CMP.

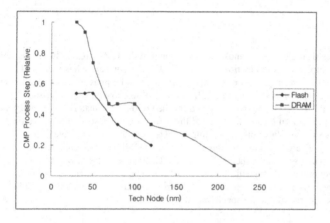

Figure 1. Increase of CMP process steps with respect to tech node.

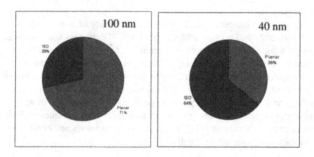

Figure 2. Increase of Isolation CMP process steps with respect to tech node.

Additionally, the CMP processes for global and local planarization have been implemented to replace conventional planarization processes since the planarity requirement for nano-lithography becomes more stringent. Figure 3 shows one of emerging CMP application for nano-lithography. As even an immersion ArF lithography technology may face a patterning limit at below 40 nm tech node, the double patterning technologies have been developed by combining a partition mask, etch, spacer deposition, and a CMP. The CMP for this spacer double patterning needs to polish the spacer materials and stop at different layers such oxide, nitride, and poly Si.[2] The successful planarity after the spacer CMP can be achieved under new selectivity condition. Until EUV lithography becomes ready and production-worthy, the CMP processes for the double patterning will increase gradually.

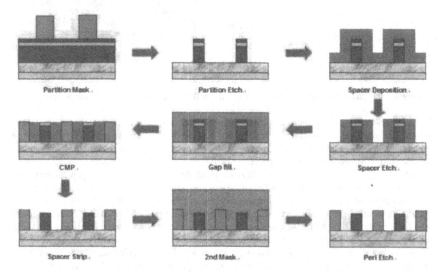

Figure 3. Application of CMP process for Spacer Double Patterning Technology. [2]

ISSUES OF ISOLATION MODULE

The characteristics of the CMP process for the flash memory are related to device structure and integration scheme. In comparison to other IC devices, the unique device structure of a flash memory is a floating gate as a memory cell unit. As shown in figure 4, a memory cell is composed of a tunnel oxide, a floating gate poly Si, an inter-poly dielectric, and a control gate. When a positive high electric field is applied to control gate during a program step, electrons are moving into the floating gate by tunneling. The charged floating gate results in the shift of threshold voltage of the cell transistor. Therefore, the threshold voltage shift is used as a signal of data storage at the memory cell. In order to write and read data on giga-scale bit memory cells, the shape and the dimension of the cell must be uniformly controlled.

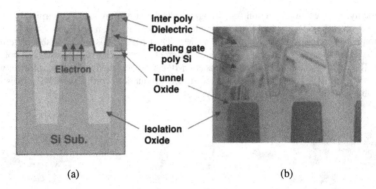

(a)	(b)

Figure 4. Schematic (a) and micrograph (b) of the cell structure of a NAND flash memory

One of critical steps of the flash memory process integration is the control of the isolation oxide height. The oxide height affects the coupling ratio and the threshold voltage of a cell transistor. A challenge in the ISO CMP for a flash memory is to achieve the formation of uniform and robust isolation oxide within both chip and wafer.

As the isolation trenches of memory devices shrink down to nano-scale, the gap-fill material was replaced from the conventional HDP oxide to more flowable solution such as SOD. The implementation of the SOD had new process requirement for the ISO CMP. Since the SOD material is relatively softer, the dishing at the SOD is worse than the HDP oxide. Therefore, the selectivity of the SOD to the hard mask nitride should be properly adjusted to minimize the dishing and erosion. The dimension of the isolation line is the same as the design rule of a nanoscale device. As the device scales down, the dimension of cell region shrinks down more than the peripheral region. Therefore, the dishing at the peripheral trench will be worse than at the cell area. Figure 5 shows the big dimension difference in between cell and peripheral trenches.

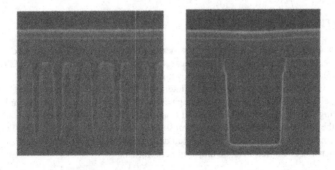

(a)	(b)

Figure 5. Micrographs showing DRAM isolation trenches filled by SOD
: (a) cell array area, (b) peripheral area

Therefore, the high selectivity slurry should be applied to minimize the dishing at the peripheral area. Since the SOD is more vulnerable to the CMP scratches, the high selectivity slurry should be formulated to minimize scratches during the CMP. The elimination and prevention of those defects must be insured by proper post CMP cleaning processes.

ISSUES IN GATE AND CAPACITOR MODULE

As the device shrinks down, the resistance of the gate electrode should be lower to meet the device requirement. The material for the gate electrode needs to be changed from tungsten to CoSi or NiSi. In order to get the uniformity of the gate electrode resistance, the uniformity of the poly Si should be obtained before the salicidation of the poly Si with Co or Ni.

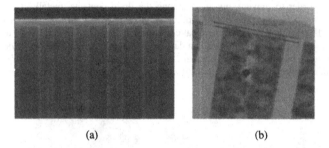

(a)   (b)

Figure 6. (a) Micrographs of Storage TiN (b) Atomic scale flatness after CMP

As shown in figure 6, the dishing at the TiN storage should be minimized to ensure the uniformity of a capacitor at nano-scale DRAM. Ru has been one of candidate electrode materials for a nano-scale DRAM. The production-worthy Ru slurry needs to be developed.

ISSUES IN CONTACT AND METALLIZATION MODULE

As the contact size shrinks down to nanoscale, the control of dishing and erosion will be critical. The planarity of underlayer CMP will affect directly the following layer uniformity. The CMP selectivity needs to be properly optimized with respect to an integration scheme and the characteristics of exposed surfaces. Figure 7 shows micrographs of scratches after a landing plug poly CMP. The scratches at the landing plug will cause bridges among the plugs and should be prevented by the optimization of the CMP slurry and process condition.

Figure 7. Micrographs of scratches after landing plug poly CMP.

As shown in figure 8, the high-aspect ratio contacts were formed by a CVD W fill and isolated by the CMP. In order to prevent any short among the drain contacts, metallic CMP residue free condition should be achieved by the optimization of the CMP process and post cleaning. During the CMP post clean, a clean chemical can attack the W plug through the seam. The contact voids can result in the failure of reading a data from a memory cell. The contact fail can be prevented by reducing hydrogen peroxide content in the chemical. As contact size shrinks down, the cleaning should be designed to minimize any damage on the contact while ensuring a residue-free surface.

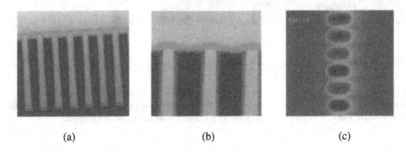

(a)   (b)   (c)

Figure 8. Micrographs showing drain tungsten contact.
(a) Drain contact arrays, (b) contacts with low dishing (c) contact voids (top view)

At the nano-scale DRAM and NAND flash memory, the Cu metallization should be implemented to ensure the speed of the data transfer. As shown in figure 9, a Cu metal line should have minimal dishing or erosion. And also, the uniformity of Cu CMP within wafer and chip should be guaranteed to meet the uniformity of the interconnect line resistance. The selectivity of Cu and barrier metal should be obtained over 100. In future, a noble barrier metal like Ru will be implemented. Up to now, the Ru CMP slurry is not ready yet since the removal rate of Ru CMP is not high enough. The Cu metal corrosion should be prohibited by proper post CMP cleaning.

Figure 9. Micrographs showing Cu metal layer with atomic scale dishing control

Figure 10 shows watermarks on W line after metal CMP. The water marks were formed during the drying process. The composition of the watermark is mainly oxide and the watermark can be prevented by the modification of drying method. The CMP post clean of the metal CMP must be done to prevent the contamination of metals. The drying after wet clean should be done without any watermarks and residues.

Figure 10. Micrographs showing watermark defects on damascene metal line

CHALLENGES IN EMERGING CMP APPLICATION

As the lateral shrinkage of the current memory devices is approaching to the physical and electrical limits, a variety of emerging solutions have been explored.[3] One feasible solution is the vertical integration of the current memory chips and the other one is the creation of new memories such as phase change memory, magnetic memory and resistive memory. In order to enable these new innovative approaches, new CMP solutions should be developed in terms of both equipments and consumables.

In order to overcome the degradation of the current memories with the tech shrinkage, a variety of new memories have been introduced and developed for the purpose of proving the production-worthiness. As the confined PRAM cell is adopted for better device performance, the GST CMP has been developed [4]. Especially, the control of the selectivity between GST and neighboring insulators will be critical since the GST material is too soft to be polished. The new CMP slurry should to be developed to ensure a good selectivity and a stable removal rate. The introduction of the new slurries in an existing consumable set must be qualified in terms of cross-contamination and reliability. Since MRAM and ReRAM are composed of quite novel materials such as ferromagnetic materials and new electrode materials, new CMP process should be developed. And also, the atom-scale dishing control and thickness uniformity must be achieved for within die and chip level device performance. As new memory structures such as SONOS and Fin-FET have been developed to overcome the degradation of device performance at nano-scale, the CMP method should be modified to be appropriate for the new integration scheme.[5] One of critical challenges in a nano-scale wafer yield management is the detection and the eradication of nano-size particles and defects. However, even a state-of-art wafer inspection equipment can not detect some nano-size defects or marks which may cause serious yield loss. Undetectable atom-size CMP residues or scratches may grow to larger defects after deposition of a thin film. The TSV scheme requires the polishing of metal layers over 10 um thickness [6]. In order to obtain the production-worthy process. the removal rate of Cu must be increased dramatically while meeting the requirements of uniformity and dishing. It is a big paradigm shift in both equipment and consumables.

CONCLUSION

There are also lots of CMP issues toward the successful nanoscale memory manufacturing. The number of the CMP processes has increased exponentially due to the complexity of integration scheme. The CMP requirements for planarization have been tougher since the planarity requirement for nano-lithography becomes more stringent. There are huge demands on the reduction of CoO since the increase of the CMP process affects the cost of manufacturing. Various new materials and integration schemes have been introduced to overcome critical issues of device performance and patterning. The new materials and integration schemes require new CMP process development. The CMP selectivity must be tuned for the various new materials for a variety of new materials and device structures. The vertical integration needs the new CMP process such as a significantly high removal rate. The dishing and erosion control are critical for the isolation of the nanoscale patterning. Even if there are many issues and challenges in the CMP for the nanoscale memory manufacturing, there will be lots of opportunities at both equipment and consumables.

REFERENCES

1. J.M. Steigerwald et. al., IDEM, San Francisco, USA (2008)
2. W.Y. Jung et. al., SPIE Advanced Lithography, San Jose, USA (2009)
3. S. Lai, IDEM, San Francisco, USA (2008)
4. Z. Liu and et. al., ICPT Proceedings, Hsinchu, Taiwan (2008)
5. S.W. Chung et. al., Symposium on VLSI Tech., Kyoto, Japan (2007)
6. Moon-Ki Jeong, The 38th Korean CMP UGM, Ansan, Korea (2008)

Tool/Process Development Such as eCMP and Low-shear CMP

Mater. Res. Soc. Symp. Proc. Vol. 1157 © 2009 Materials Research Society 1157-E12-02

Optimization of Material Removal Efficiency in Low Pressure CMP

Bozkaya Dinçer, Müftü Sinan

Department of Mechanical Engineering, Northeastern University
360 Huntington Ave, Boston, Massachusetts, 02115

ABSTRACT

In chemical-mechanical polishing (CMP) the material removal efficiency (MRE) can be defined as the fraction of the total pressure distributed on the abrasives, and it depends on the interplay between the direct contact of the pad-to-wafer, and the contact of the abrasives with the wafer. The MRE can be increased by minimizing pad-wafer direct contact, as this is not likely to help material removal, significantly. The objective of this work is to investigate parameters that control MRE. This may be especially important for low-pressure CMP used in the polishing of (ultra-low-k) ULK dielectric materials. The optimization of CMP parameters to maximize the MRE is described by modeling the contact interactions between pad, abrasives and wafer. A relationship for optimal abrasive concentration is presented for the external load values that mark the transition from pure pad-wafer-abrasive contact to mixed contact (combination of pad-wafer-abrasive and pad-wafer contacts) and for given pad porosity and pad surface parameters.

INTRODUCTION

In this work the interactions between the pad, the wafer and the abrasive particles are modeled at different contact scales. At the heart of this model are relationships, obtained by finite element modeling, for *three-body contact force* due to the contact of the pad, the wafer and the abrasive particles (Fig. 1a), and the *two-body contact force* between the pad and the wafer. The results of this *single particle* (SP) *model* are used in a *multi-particle* (MP) *model* describing the contact of a flat wafer with a flat deformable pad (Fig. 1b). The curve fit relationships from the multi-particle model are used to represent the deformation of a *single asperity* in the presence of interfacial particle [1]. These relationships are in turn used in determining the contact conditions between a rough-pad and a smooth wafer in the *multi-asperity model*.

Polishing pads are typically made of porous polymeric materials; and, the elastic modulus of the porous pad E_p, is typically ¼-to-⅓ of the elastic modulus of the polymer's solid elastic modulus, E_s. The pad porosity is introduced in the model by considering the relative size of the abrasives with respect to the pad asperity. We assume that the pad's local interactions with the abrasives are dominated by the solid pad elastic modulus E_s. On the other hand, the multi-asperity contact of the pad with the wafer is influenced by the pad's porous elastic modulus, E_p.

In typical CMP slurries the abrasive particle size is typically variable and on the order of 20-100 nm. In this work, by assuming a constant abrasive size we are able derive a closed form relationship for maximizing the material removal efficiency.

CONTACT MODELS

The SP and MP contact models, [1] are summarized for the case of constant particle size.

Single Particle Contact

The SP contact model (Fig. 1a) analyzes the contact of a single particle with pad and wafer using a finite element model. In this work the pad is modeled as a hyperelastic material, with a two-parameter Mooney-Rivlin model, with coefficients $a_{10} = 0.5$ MPa and $a_{01} = 0.5$ MPa, yielding $E_s = 6$ MPa [2]. The rigid wafer and particle are penetrated into the deformable pad incrementally, and the finite element model is solved for the deformation of the pad for each increment. The particle contact regime prevails at small penetrations, where the entire load is carried by pad-to-particle (particle) contact. In the mixed contact regime, the load is shared by particle contact and pad-to-wafer (direct) contact (Fig. 1a). In this regime, the penetration is measured with respect to displacement of deformable surface, δ_d, as depicted in Fig. 1a, which is non-dimensionalized as $\varepsilon_s = \delta_d / t_s$, with respect to thickness of deformable medium, t_s. The contact force acting on the particle (*particle contact force*) $f_p^{m^*}$ ($= f_p^m / E_s r_p^2$) is described by the following (curve-fit) relationship,

$$f_p^{m^*}(\varepsilon_s) = \begin{cases} 5.4(\varepsilon_s)^{0.57} + 3.12, & \text{for } 0 < \varepsilon_s < 0.05 \\ 11.1(\varepsilon_s - 0.05)^{0.90} + 4.10, & \text{for } 0.05 < \varepsilon_s < 0.2 \\ 40.94(\varepsilon_s - 0.2)^2 + 13.14(\varepsilon_s - 0.2) + 6.11, & \text{for } 0.2 < \varepsilon_s < 0.45 \end{cases} \quad (1)$$

where r_p is the particle radius. The *influence radius* r_i^* ($= r_i / r_p$), which is used to quantify the non-contact region around a particle (Fig. 1a) is expressed with the following relationship,

$$r_i^*(\varepsilon_s) = 1.52(\varepsilon_s)^{-0.45} \quad \text{for } 0 < \varepsilon_s < 0.45 . \quad (2)$$

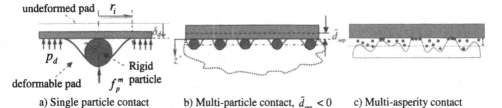

undeformed pad r_i		
p_d		$\tilde{d}_{sep}$
deformable pad f_p^m Rigid particle		
a) Single particle contact	b) Multi-particle contact, $\tilde{d}_{sep} < 0$	c) Multi-asperity contact

Fig. 1: Contact models at different scales

Multi-Particle Contact

In this section, multi-particle contact illustrated in Fig. 1b is studied. The number of particles per unit slurry volume n_v, can be obtained by using the particle concentration by weight η_w and the densities of the slurry ρ_s and the particle material ρ_p as follows,

$$n_v^* = \frac{3}{4\pi} \frac{\rho_s}{\rho_p} \eta_w. \tag{3}$$

where $n_v^* = n_v r_p^3$ is the dimensionless volumetric particle concentration. All particles in the slurry become active when the wafer-to pad separation distance d_{sep} equals the particle diameter, $d_{sep} = 2r_p$, due to the constant particle diameter assumption. The number of active particles n_a^{mp} over a surface area A, can be found by using this separation as $n_a^{mp} = n_v A d_{sep}$. By using the non-dimensional variables $d_{sep}^* = d_{sep} / r_p$, $A^* = A / r_p^2$ the non-dimensional number of active particles per area is expressed as follows $n_a^{mp*} = 2n_v^*$ with $n_a^{mp*} (= n_a^{mp} r_p^2)$.

The direct contact area between the pad and the wafer, A_d^{mp*}, can be obtained $A_d^{mp*} = 1 - A_i^{mp*}$. The influence contact area A_i^{mp*} can be found by the summation of influence contact area of each particle, πr_i^{*2} over all active particles as $n_a^{mp*} \pi r_i^{*2}$. Using the number of active particles n_a^{mp*}, the influence contact area A_i^{mp*} becomes,

$$A_i^{mp*} = 2\pi r_i^{*2} n_v^* \tag{4}$$

Substituting the volumetric particle concentration n_v^* (Eqn (3)) and influence radius r_i^* (Eqn (2)) into Eqn (4), the direct contact area A_d^{mp*} can be expressed as a function of average compressive strain ε_p,

135

$$A_d^{mp^*} = 1 - 3.47 \frac{\rho_s}{\rho_p} \frac{\eta_w}{(-\varepsilon_p)^{9/10}} . \tag{5}$$

Direct contact of surfaces occurs when the direct contact area becomes $A_d^{mp^*} \geq 0$. The critical average compressive strain ε_p^m, at which direct pad-to-wafer contact starts to occur, is found, by setting the direct contact area, $A_d^{mp^*} = 0$ in Eqn (5), as follows,

$$\varepsilon_p^m = -3.98 \left(\frac{\rho_s \eta_w}{\rho_p} \right)^{10/9} . \tag{6}$$

The contact pressure at this critical condition is computed by using $\varepsilon_p = -\varepsilon_p^m$ in Eqn (1) and the number of active particles $n_a^{mp^*}$ as follows,

$$p_p^{mp-c^*} = \frac{p_p^{mp-c}}{E_s} = 2n_v^* f_p^{m^*} \left(-\varepsilon_p^m \right) . \tag{7}$$

Optimization Equation

The mean real contact pressure due to contact of a rough pad and a smooth wafer (Fig. 1c) is calculated using Greenwood and Williamson [3] multi-asperity model, and assuming that the asperity summit heights are distributed according to an exponential distribution. In this case, the mean real contact pressure p_c^m is constant with respect to applied pressure and is a function of pad surface parameters as follows [3],

$$p_c^m \approx 0.39 E_p \sqrt{\frac{\sigma_s}{R_s}} \tag{8}$$

where σ_s is the standard deviation and R_s is the mean radius of the summits for pad roughness, and E_p is the elastic modulus of for porous pad. Considering the local contact pressure in Eqn (7) and mean real contact pressure to be equal, $p_p^{mp-c^*} E_s = p_c^m$, the following relationship is obtained,

$$0.82 \sqrt{\frac{\sigma_s}{R_s}} \frac{\rho_p}{\rho_s} = \eta_w \frac{E_s}{E_p} f_p^{m^*} \left(-\varepsilon_p^m \right) . \tag{9}$$

Note that Eqn (9) is obtained for the optimal distribution of the contact pressure on the abrasives, and thus represents the optimal MRE. The following three parameter groups (σ_s/R_s), (ρ_p/ρ_s), (E_s/E_p), and the particle concentarion η_w are considered as the problem parameters.

RESULTS AND DISCUSSION

Optimal particle concentrations η_w^{opt} were determined from Eqn (9) for a range of pad roughness values (σ_s/R_s), for alumina and silica particles with $\rho_p/\rho_s = 3.7$ and 2.5, respectively. The pad surface parameters and pad elastic modulus were varied in the range $0.01 < \sigma_s/R_s < 1$ and $1 < E_s/E_p < 4$, respectively. Note that for a typical CMP pad the ratio $\sigma_s/R_s \sim 0.1$ ($\sigma_s = 5\mu m$ and $R_s = 50 \ \mu m$) and $E_s/E_p \sim 3$ (for 40% porosity).

Fig. 2 shows that optimum particle concentration η_w^{opt} increases with increased pad roughness (larger σ_s/R_s) and/or decreased pad porosity (smaller E_s/E_p). As σ_s/R_s increases, the mean real contact pressure at the tip of asperities becomes greater; as a result, more particles are required to prevent direct contact at pad-wafer interface. Higher solid pad elastic modulus E_s decreases the tendency of particles to be embedded in the pad. As the porous pad modulus E_p increases, the real contact area between pad and wafer becomes smaller, thus increasing the mean real contact pressure and the number of particles (or η_w^{opt}) to prevent direct contact. As a result of these two opposing effects, optimum particle concentration η_w^{opt} is not a function of solid E_s or porous pad elastic modulus E_p, but the ratio E_s/E_p. The optimum particle concentration is also found to be greater in the case of alumina particles since the mass density of alumina particle is greater than silica particles resulting in a greater number of active particles for a given particle density by weight ratio η_w^{opt}.

CMP experiments by Guo et al. [4] and Bielmann et al. [5] with alumina-slurry illustrates that there exists a saturation particle concentration above which MRR remains constant as particle concentration increases (Fig. 3). The saturation particle concentration seen in experiments corresponds to the optimum particle concentration, which marks the transition from mixed to particle contact regime. The transition from mixed to particle contact regime is found to occur at particle concentration $\eta_w^{opt} = 5\text{-}8\%$ in experiments which agree with model predictions for a porous pad, $E_s/E_p \sim 3$ and typical pad surface parameters $\sigma_s/R_s \sim 0.1$.

CONCLUSIONS

This study shows that the material removal efficiency may be optimized by considering the pad surface topography, pad porosity and particle concentration. The optimization may be used as a tool for determining the conditions to decrease the stresses in ultra-low-k (ULK) dielectric materials in CMP.

REFERENCES

1. Bozkaya D., and Müftü S., 2008. The effects of interfacial particles on the contact of an elastic sphere with a rigid flat surface. ASME Journal of Tribology, 130, 4, 041401.
2. Kim A.T., Seok J., Tichy J.A., Cale T.S., 2003. A multiscale elastohydrodynamic contact model for CMP. J Electrochem Soc, 150, 570-576.
3. Greenwood J. A., Williamson J. B. P., 1966. Contact of nominally flat surfaces. Proceedings of the Royal Society of London, Series A295, 300-319.
4. Guo L., Subramanian R.S., 2004. Mechanical removal in CMP of copper using alumina abrasives. J Electrochem Soc, 151, 104-108.
5. Bielmann M., Mahajan U., Singh R. K., 2002. Effect of particle size during tungsten chemical mechanical polishing. Electrochem Solid State Letters, 2, 401-403.

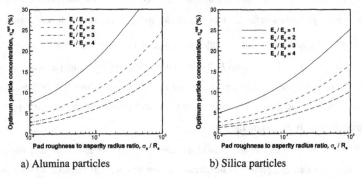

a) Alumina particles   b) Silica particles

Fig. 2: The optimum particle concentration η_w^{opt} of a) alumina and b) silica particles as a function of pad topography σ_s / R_s for different pad porosity E_s / E_p.

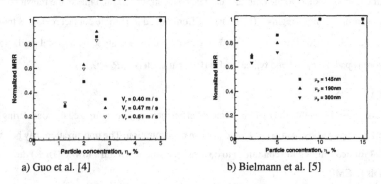

a) Guo et al. [4]   b) Bielmann et al. [5]

Fig. 3: Experimental observations for saturation of MRR at high particle concentration η_w. MRR is normalized with respect to the maximum MRR observed in experiments.

Mater. Res. Soc. Symp. Proc. Vol. 1157 © 2009 Materials Research Society   1157.F12-03

Role of Phosphoric Acid in Copper Electrochemical Mechanical Planarization Slurries

Serdar Aksu
SoloPower Inc., 5981 Optical Court, San Jose, CA 95138, U.S.A.

ABSTRACT

In this paper, the electrochemical behavior of copper in aqueous solutions containing phosphoric acid (H_3PO_4) is investigated to elucidate the role of H_3PO_4 in the Cu ECMP slurries. Aqueous solubility and potential-pH diagrams were constructed for copper-phosphate-water system. Good correlations were found between the diagrams and the experimental polarization data. It was found that H_3PO_4 might not able to sufficiently increase the solubility of copper alone. A complexing agent is needed to ensure the high solubility of copper, especially as the slurry pH and dissolved copper concentration increase. Specific conductance measurements revealed that phosphoric acid was the key constituent responsible for increasing the conductivity of the ECMP electrolyte. *In situ* electrochemical polarization experiments showed that the planarization mechanism during the ECMP process was similar to that observed in conventional copper CMP.

INTRODUCTION

Planarization of copper damascene structures is becoming increasingly challenging due to the ongoing reduction of feature sizes, along with the associated high packing density, use of low-k dielectrics and growing number of interconnect layers in silicon integrated circuits [1, 2]. Electrochemical mechanical planarization, ECMP is a technique, which can increase the planarization efficiency and the throughput at lower down force and lower defectivity [3,4]. In contrast to conventional CMP process, where chemical oxidizers are primarily responsible for controlling the polishing rate, in the ECMP process applied current density is used to control the polishing rate. ECMP offers the high planarization efficiency, which is insensitive to the pattern density [5]. Unlike the conventional electropolishing process, which uses inorganic acids that dissolve copper uniformly from the entire wafer surface, ECMP electrolyte produces a soft protective layer that is removed easily by the pad abrasion from the protruded areas. On the other hand, the mechanical action leaves the soft protective film intact in the recessed areas, thus channeling more current to the abraded areas. Selection of correct pad stiffness is critical in ECMP process to ensure high planarization efficiency. A soft pad would bend and remove the protective surface layer in the recessed areas while a stiff pad would not respond well to the low pressure in the ECMP process [5].

Planarization of the copper film depends highly on the electrolyte chemistry in the ECMP process. For best results the protective surface film must be soft and easily abraded when contacted by the pad while it should remain intact in the recessed areas. The main chemical constituents of ECMP slurries could be listed as complexing agents, passivating agents (corrosion inhibitors), pH buffers and inorganic acids [6]. Oxidizers, such as H_2O_2, are sometimes also added to enhance the etching from the abraded regions during the planarization process. In essence, the chemical composition of copper ECMP slurries is similar to those used in the Cu CMP process. Complexing agents such as citric acid, oxalic acid, ethylenediamine, glycine, malic acid and ammonium citrate are added into the CMP and ECMP slurries to increase the driving force for dissolution through complexation of the dissolved copper ions [2, 6]. Common passivating agents that could be used in both CMP and ECMP slurries include benzotriazole [6,7], 5-aminotetrazole [8] and 5-phenyl-1-H-tetrazole [1]. These agents protect the recessed sections of copper from the anodic dissolution during the planarization process. As in the case of Cu CMP slurries, buffering agents should also be included to maintain the pH of ECMP slurry at a desired value [6].

ECMP slurries also employ inorganic acids, such as phosphoric or sulfuric acid and their derivatives. These inorganic acids are expected to raise the conductivity of the electrolyte and promote dissolution of copper from the abraded sections during the dynamic polishing conditions. However, there is very limited amount of work on the specific functions of inorganic acids in the ECMP slurries [9]. In this paper, the role of phosphoric acid in the Cu ECMP slurries is investigated in detail. Effect of phosphoric acid on the dissolution of copper is examined by establishing the aqueous solubility and potential-pH diagrams for copper-phosphate-water system and comparing their predictions with potentiodynamic polarization experiments. The effect of phosphoric acid in increasing the ECMP electrolyte conductivity is delineated. The planarization mechanism during the Cu ECMP process in slurries containing phosphoric acid was defined using the experimental data from *in situ* electrochemical polarization experiments.

EXPERIMENTAL

Potentiodynamic polarization studies were conducted using the electrode system shown in Figure 1, which included a modified rotating disk electrode system (Pine Instruments Co.) and a computer-controlled potentiostat/galvanostat (Princeton Applied Research, model 263-A). The copper working electrode was constructed using an oxygen-free copper rod, 2.54 cm in diameter and 4 cm tall, which could be screwed into the rotator. The surface of the copper rod was insulated to leave an unsealed copper disk with about 0.3 mm in height at the bottom. All the potentiodynamic polarization curves for rotating copper electrode were obtained at 1000 rpm and at scan rate of 2 mV/s. A saturated calomel electrode, SCE, was used as a reference electrode while the measured potentials were reported here with respect to the standard hydrogen electrode, SHE. The counter electrode was made from stainless steel. The concentration of H_3PO_4 was 0.5 M in all experiments. In addition, ECMP slurry contained 0.1 M citric acid, 0.25 M BTA and 0.4 wt % colloidal silica particles (Rohm and Haas) and had a pH value of 4. An IC1000 pad from Rohm and Haas was used in experiments with abrasion. In the absence of abrasion, the copper electrode was about 5 mm above the pad surface. A high-precision balance (Ohaus Co.) was used to measure the applied force. The pressure, which is the weight applied per surface area of the pad, was maintained at 1 psi in all abrasion experiments. During abrasion, the pad masked some portion of the copper working electrode area. The current densities reported under abrasion were calculated using the unmasked portion of the copper surface. AYSI 30 model handheld conductivity meter was used in the conductivity measurements.

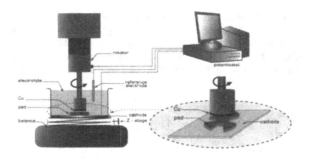

Figure 1. Experimental set-up used in electrochemical experiments

RESULTS AND DISCUSSION

Phosphoric acid can exist in aqueous solutions in five different forms, namely, $H_4PO_4^+$, H_3PO_4, $H_2PO_4^-$, HPO_4^{2-}, PO_4^{3-}. The equilibria between these species may be depicted as [10]:

$$\underset{H_4PO_4^+}{} \overset{pK_{a1}=0.0}{\leftrightarrow} \underset{H_3PO_4}{} \overset{pK_{a2}=2.148}{\leftrightarrow} \underset{H_2PO_4^-}{} \overset{pK_{a3}=7.198}{\leftrightarrow} \underset{HPO_4^{2-}}{} \overset{pK_{a4}=12.375}{\leftrightarrow} \underset{PO_4^{3-}}{} \quad (1)$$

Phosphoric acid could form soluble complexes with both cuprous and cupric ions. CuH_2PO_4, $Cu(H_2PO_4)_2^-$ and $CuH_3(PO_4)_2^{2-}$ are the aqueous Cu(I) phosphate complexes. Aqueous Cu(II) phosphate complexes include $CuHPO_4$, $CuH_2PO_4^+$, $Cu(H_2PO_4)_2$, $CuH_2(PO_4)_2^{2-}$, $CuH_3(PO_4)_2^-$ [10]. $Cu_3(PO_4)_2$ was considered as the principal cupric-phosphate solid in the present study [11].

Figure 2a shows the potential-pH diagram constructed at $\{OP_T\}=0.5$ and $\{Cu_T\}=10^{-3}$ using the most recent thermodyanamical data [10,11] for the copper-phosphate-water system at 25 °C and 1 atm. The potential-pH diagram shows that there is an increase in solubility range of copper due to the formation of aqueous copper-phosphate species. At pH values below the CuO stability domain, Cu^{2+}, $CuH_2PO_4^+$, $Cu(H_2PO_4)_2$, $CuH_3(PO_4)_2^-$ and $CuH_2(PO_4)_2^{2-}$ are the predominant aqueous species in the order of increasing pH. Cu_2O predominance region start at a pH of about 4.8 as a thin band whereas CuO is stable at pH values above 8.7 under these conditions. At highly alkaline pH values copper becomes soluble due to the formation of CuO_2^{2-}. These finding are in agreement with the aqueous solubility diagram shown in Figure 2b, which displays the influence of pH and $\{Cu_T\}$ on the solubility of $Cu_3(PO_4)_2$ and CuO at $\{OP_T\}=0.5$. The exact solubility curve in Figure 2b indicates that $\{Cu_T\}_{min}$ for the formation of $Cu_3(PO_4)_2$ is about 1.2×10^{-3} at pH=5.1 for $\{OP_T\}=0.5$. At pH values between about 4.3 to 6, $Cu_3(PO_4)_2$ is expected to form at $\{Cu_T\}$ values above 2×10^{-3}. On the other hand, onset of $Cu_3(PO_4)_2$ formation is about 3×10^{-3} at pH=4.

Figure 3a shows potentiodynamic polarization curves of copper in 0.5 M H_3PO_4 at pH values of 4 and 6. The behavior shown in Figure 3a is fairly consistent with the potential-pH diagram shown in Figure 2a. The polarization curve obtained at pH 4 shows active dissolution behavior, with no evidence of passivation. On the other hand, the anodic portion of the polarization curve at pH 6 shows that passivation of copper surface at potentials above 600 mV. Figure 2a suggest that Cu_2O can form at pH value of about 4.8 at $\{OP_T\}=0.5$ when $\{Cu_T\}$ reaches to a value of 10^-3. Therefore, the passivation behavior observed at pH 6 might be due to formation of Cu_2O on the copper surface if dissolved copper concentration reached to 10^{-3} M at anodic potentials above 600 mV. According to the aqueous solubility diagram shown in Figure 2b, formation of $Cu_3(PO_4)_2$ might also contribute to the passivation behavior if $\{Cu_T\}$ exceeds $2x10^{-3}$.

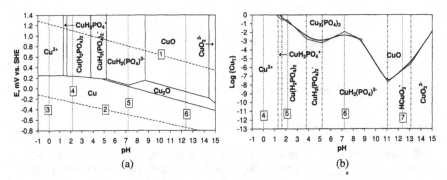

Figure 2. (a) Potential-pH diagram at $\{OP_T\}=0.5$ and $\{Cu_T\}=10^{-3}$ (b) aqueous solubility diagram at $\{OP_T\}=0.5$ for the copper-phosphate-water system at 25 °C and 1 atm.

ECMP slurry studied in this paper contained 0.25 M citric acid as the complexing agent to further increase the solubility of copper. The potential-pH diagrams constructed by Tamilmani and co-workers for copper-citric acid-water system showed that the solubility domain of copper can be increased significantly with addition of citric acid [12]. Anodic polarization curve in the absence of BTA shown in Figure 3b indicates that the electrochemical behavior of copper did not change significantly when 0.1 M citric acid was added to the electrolyte containing 0.5 M H_3PO_4 at pH 4. However, addition of same amount of citric acid prevented passivation of the copper surface at pH 6. An active dissolution behavior was observed from the anodic polarization curve at this pH (not shown here). Figure 3b shows that addition of 0.25 M BTA as the passivating agent increased the open circuit potential (E_{OCP}) and reduced the dissolution rate considerably due to the formation of a passive film on the copper surface. When copper surface was abraded, there was a significant drop in E_{OCP}, indicating that the passivation layer was completely removed. The open circuit potential obtained under abrasion was much lower than that observed in the absence of passivating agent. This suggests that the mechanical action does not only remove the passivation layer but also activates the copper dissolution due to the formation of fresh surface and enhanced mass transport. The planarization mechanism during the ECMP process is quite similar to that observed in conventional copper CMP [13]. In the presence of a passivating agent, a protective passivation layer is formed on the copper surface. The protective film is removed by the mechanical action, predominantly at the protruding regions. An aggravated active anodic dissolution takes place at the protruding sections as the copper surface rendered unprotected there. In contrast, recessed regions of the copper surface are initially passivated and undergo little mechanical damage and material removal until the entire surface has been planarized. Similar results to those displayed in Figure 3b were also obtained at pH=6 (not shown). Therefore, high planarization and removal rates can be achieved at the pH range of 4 to 6.

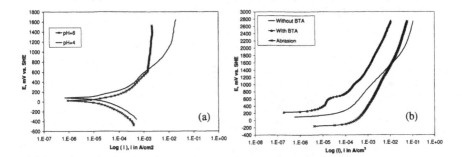

Figure 3. (a) Polarization behavior of copper in 0.5 M H_3PO_4 at pH values of 4 and 6 (b) Anodic polarization curve showing the effects of addition BTA and abrading the copper surface at pH 4 (Scan rate is 2 mV/sec and rotational speed is 1000 rpm for in all experiments).

In addition to increasing the solubility of copper, phosphoric acid offers several other desirable properties in ECMP slurries. It is a non-toxic and non-volatile acid, which does not act as an oxidant. Figure 4 shows the change in the specific conductance with addition of each chemical constituent of the ECMP slurry used in the present study. It should be noted that the specific conductance of de-ionized water is only 3.6×10^{-6} S. As can be concluded from the comparison of Figure 4a and 4b, phosphoric acid is the key component for increasing the conductivity of the ECMP electrolyte. Figure 4b shows that when phosphoric acid is excluded the same formulation of the ECMP electrolyte can provide only fifth of the electrolytic conductivity. A highly conductive electrolyte in the ECMP slurry is crucial to minimize the solution phase ohmic potential drop. Uniformity of the electrochemical current distribution over the wafer surface might be negatively impacted if the limiting current range is masked by ohmic potential drops at low electrolyte conductances [9].

CONCLUSIONS

The electrochemical behavior of copper in aqueous solutions containing phosphoric acid (H_3PO_4) was investigated by establishing the aqueous solubility and potential-pH diagrams for copper-phosphate-water system. The predictions of the diagrams were in agreement with the experimental data from potentiodynamic polarization study. It was determined that solubility of copper might not be increased sufficiently with H_3PO_4 alone. Especially at higher pH values and copper dissolution rates, a complexing agent is needed to ensure the high solubility of copper. Phosphoric acid was found to be the key constituent for increasing the conductivity of the ECMP electrolyte. *In situ* electrochemical polarization experiments indicated that the planarization mechanisms for CMP and ECMP processes were similar. In the presence of BTA, a passivation layer is formed on the copper surface, which minimized the material removal from the recessed regions. However, protective film is removed by the mechanical action, predominantly at the protruding regions, leading to an aggravated active anodic dissolution. This differential in the material removal between recessed and protruding regions is responsible for producing an increasingly planarized surface.

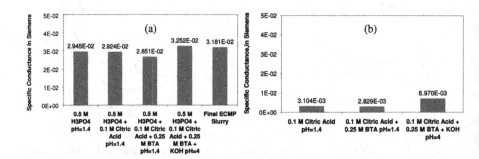

Figure 4. Specific conductance (a) in the presence and (b) in the absence of phosphoric acid.

REFERENCES

1. K. W. Chen, Y. L. Wang, C. P. Liu, L. Chang, F. Y. Li, Thin Solid Films, 498, 50 (2006).
2. Y. Hong, D. Roy, S. V. Babu, Electrochem. Solid-State Lett., 8, G297 (2005).
3. A.Tripathi, C. Burkhard, I I. Suni, Y. Li, F. Doniat, A. Barajas, and J. McAndrew, J. Electrochem. Soc, 155, H918 (2008).
4. A. C. West, I. Shao, H. Deligianni, J. Electrochem. Soc., 152, C652 (2005).
5. S. Aksu, I. Emesh, C. Uzoh and B. Basol, ECS Trans. 2, 417 (2006).
6. F. Q. Liu, S. D. Tsai, Y. Hu, S. S. Neo, Y. Wang, A. Duboust, L-Y Chen, US Patent No. 7,128,825 (October 31, 2006).
7. M. T. Wang, M. S. Tsai, C. Liu, W. T. Tseng, T. C. Chang, L. J. Chen, M. C. Chen, Thin Solid Films 308, 518 (1997).
8. J-W. Lee, M-C Kang, J. J. Kim, J. Electrochem. Soc, 152, C827 (2005).
9. K. G. Shattuck, J-Y Lin, P. Cojocaru, A. C. West, Electrochim. Acta, 53, 8211 (2008).
10. R. M. Smith and A. E. Martell, NIST Critically Selected Stability Constants of Metal Complexes Database Software, Version 8.0, May 2004.
11. J. W. Ball and D. K. Nordstrom, WATEQ4F Database, U. S. Geological Survey, Menlo Park, California (1991).
12 . S. Tamilmani, W Huang, S. Raghavan, and R. Small, J. Electrochem. Soc., 149, G638 (2002).
13. S. Aksu and F. M. Doyle, J. Electrochem. Soc., 149, G352 (2002).

**Advanced CMP Process
Control Techniques**

Mater. Res. Soc. Symp. Proc. Vol. 1157 © 2009 Materials Research Society   1157-F13-07

CMP for High Mobility Strained Si/Ge Channels

Kentarou Sawano[1], Yasuhiro Shiraki[1] and Kiyokazu Nakagawa[2]
[1] Advanced Research Laboratories, Tokyo City University, 1-28-1 Tamazutsumi, Setagaya-ku, Tokyo 158-8557, Japan
[2] Center for Crystal Science and Technology, University of Yamanashi, Miyamae-cho 7, Kofu, Yamanashi 400-8511, Japan

ABSTRACT

Chemical mechanical planarization (CMP) technique has been developed for the planarization of SiGe strain relaxed buffer layers (virtual substrates). CMP successfully eliminated the surface roughness arising inevitably on the SiGe buffer layer and ultrasmooth surface with RMS roughness less than 0.5 nm was obtained. The planarized SiGe virtual substrates were applied to strained-Si and strained-Ge channel structures and significant mobility enhancements owing to the planarization were demonstrated. These results indicated that the CMP is essential technology for next generation high performance Si/Ge based devices.

INTRODUCTION

Channel strain engineering has become an essential technology as a performance booster for the continued growth of future complementary metal oxide semiconductor (CMOS) since the introduction of strain into the channel can offer significant mobility enhancement owing to highly altered band structures. Among the variety of strain-inducing techniques developed, process-induced local strain has recently achieved mobility enhancement [1] and is currently the mainstream technology. In contrast, since there is 4.2% lattice mismatch between Si and Ge, Si/Ge heteroepitaxy can introduce biaxial strain into channels across a whole wafer, which is called the global strain technique. This technique has many superior advantages over the local strain technique, for example, a much larger amount of strain can be induced with much higher stability and uniformity and strain controllability is higher thanks to its independence from device processes. The globally strained channels such as strained-Si and strained-Ge are generally formed on strain relaxed SiGe buffer layers (called SiGe virtual substrates). In such structures, a quality of the SiGe virtual substrate is very crucial to obtain high carrier mobility in the strained channel since SiGe virtual substrate quality is inherited directly to the overgrown channel quality. To obtain high quality SiGe virtual substrates, various growth methods such as the compositionally graded buffer method [2,3] and the low temperature buffer method [4,5] have been developed so far. One critical problem, however, is a very large surface roughness arising on the SiGe virtual substrate. So-called crosshatch pattern appears on the surface, irrespective of growth methods. So far, much efforts have been made to reduce the roughness by means of adjusting growth parameters, but failed to realize SiGe virtual substrates with an ultra smooth surface comparable to that of a Si substrate, that is, root mean square (RMS) surface roughness of less than 1 nm. Since strain field coming from the underlying misfit dislocation arrays is responsible for this roughness formation [6,7], it seems essentially impossible to fabricate SiGe virtual substrates with ultra smooth surface only through epitaxial growth. It is well recognized that the graded buffer method is one of the most sophisticated and useful methods for the purpose of reducing threading dislocation density, but it is quite difficult with this method to suppress the surface roughness less than 10 nm as well as other growth methods. It is, therefore, considered that the most appropriate way to obtain the high quality virtual substrate is combination of the graded buffer technique that can offer low threading dislocation density and subsequent surface planarization via some other technique than crystal growth.

Chemical mechanical polishing (CMP), which is established technology for Si wafer production, is considered to be highly suitable for the SiGe planarization, and has been employed by several groups [8-13]. Currently, CMP has become quite standard technology for planarization of the SiGe virtual substrate, including SiGe-on-insulator (SGOI) wafer fabrication, where CMP is indispensable procedure for wafer bonding. Here we briefly review our developments of SiGe CMP technique [14,15] and its application to strained channel structures [16,17].

EXPERIMENT

SiGe strain-relaxed buffer layers (SiGe virtual substrates) with Ge compositions from 25 to 65 % were grown on Si(100) substrates with gas source (GS) molecular beam epitaxy (MBE) utilizing the conventional graded buffer technique. The graded buffer layers consisted of compositionally step-graded SiGe layers with thicknesses from 300 nm to 2 μm and uniform composition SiGe layers with 1 μm thickness. The growth temperatures were 700 ~ 800°C. The low-temperature (LT) buffer technique was also employed with solid source (SS) MBE. The LT buffer layers consisted of a 50-nm-thick Si layer grown at 400 °C and 500-nm-thick SiGe grown at 600 °C.

The CMP of the grown SiGe buffer layers was performed with the standard Si wafer final polishing procedure. As a polishing slurry, we used GLANZOX 3000 (Fujimi Incorporated) that contains colloidal silica particles (9.1 wt %) with the average diameter of 70 nm in NH_4OH solution (pH 11). BELLATRIX K0024 (Fujimi Incorporated) was used as a polishing pad and the polishing pressure, by which the wafer was pushed against the polishing pad, was typically 150 g/cm^2. Only for the purpose of investigating polishing rates, other slurry GLANZOX 3900 which contains colloidal silica particles (9.1 wt %) with the average diameter of 35 nm in NH_4OH solution (pH 11) and other polishing pressure (450 g/cm^2) were also utilized. The polished thickness, which was estimated by scanning electron microscope (SEM) measurements of the cross-section of the polished samples, was varied from 0 to 200 nm in order to investigate the polished thickness dependence of the surface roughness.

A post-CMP cleaning, which is a critical issue for the fabrication of high-quality device structures on the planarized SiGe virtual substrates, was carried out by the following procedures. First, in order to remove the silica particles, the wafers were immersed in $NH_4OH + H_2O_2 + H_2O$ (2:9:150) solution for 10 minutes. The cleaning temperature was varied from 20 to 75 °C to investigate the temperature dependence of the surface roughness. The wafers were subsequently rinsed with de-ionized water for 10 minutes, and soaked in HF (0.5 %) for 30 seconds at room temperature to remove SiO_2 on the surface together with particles in SiO_2. Methanesulfonic acid as a surfactant was added to HF in order to prevent the particles from sticking to the surface again. It was confirmed by AFM characterization that the silica particles, which covered the as-planarized surface as seen in Fig. 1 (a), were completely removed through this cleaning procedure (Fig. 1 (b)).

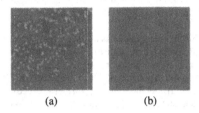

(a)   (b)

Fig. 1. AFM images of the polished surface (a) before and (b) after cleaning.

RESULTS AND DISCUSSIONS

CMP and post cleaning of SiGe virtual subsrtrates

Polishing rates were firstly estimated by measuring polished thicknesses with various polishing times, and dependencies of the polishing rates on the Ge concentration of SiGe, the size of silica particles in the slurry and the polishing pressure were investigated, which is summarized in Fig. 2. It is found that the polishing rates for $Si_{0.7}Ge_{0.3}$ increase largely with the polishing pressure for the both slurry and that the slurry with the smaller particles has the slower polishing rate. It is also found that the polishing rate for $Si_{0.4}Ge_{0.6}$ is larger than that for $Si_{0.7}Ge_{0.3}$, implying that the polishing rate is highly dependent on the Ge concentration in SiGe.

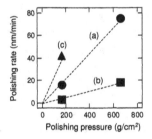

Figure 2. Polishing rates for (a) $Si_{0.7}Ge_{0.3}$ by the slurry with 70 nm silica particles, (b) $Si_{0.7}Ge_{0.3}$ by the slurry with 35 nm silica particles and (c) $Si_{0.4}Ge_{0.6}$ by the slurry with 70 nm silica particles, as a function of polishing pressure.

Figures 3 show the surface AFM images of the $Si_{0.7}Ge_{0.3}$ virtual substrate before and after CMP. The AFM scan range was 10 μm × 10 μm. The root mean square (RMS) of the surface roughness is larger than 20 nm before CMP (Fig.3 (a)). After 30 seconds polishing (Fig.3 (b)), the RMS roughness decreased to 12 nm. The upper regions of surface undulations are selectively polished. After polishing for 10 minutes (Fig.3 (c)), ultra smooth surface was realized as shown in this figure. The RMS roughness was less than 0.5 nm corresponding to several atomic layers.

Figures 4 shows power spectrum densities (PSD) obtained from the AFM measurements with scan ranges of 10×10 μm^2 and 50×50 μm^2. It is found that not only long-ranged roughness (> 1 μm) clearly seen in the AFM images (Fig. 3) but also short-ranged roughness (< 1 μm) is markedly reduced by CMP and the obtained PSD after CMP is very close to that of the Si substrate.

Figure 5 summarizes the RMS roughness values as a function of the polished thickness. The polished thickness was estimated both from the polishing times and the polishing rate (20 nm/min). It is found that the roughness rapidly decreases down to several nm after about 50 nm polishing. This polished thickness roughly corresponds to the peak-to-valley of the as-grown surface roughness as seen in Fig. 3 (a). As the polishing further proceeds, the roughness decreases gradually and finally becomes less than 0.5 nm after 200 nm polishing.

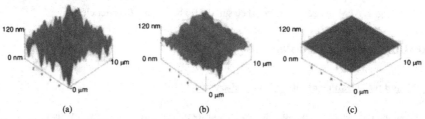

(a)   (b)   (c)

Figure 3. Surface AFM images of the $Si_{0.7}Ge_{0.3}$ virtual substrate (a) before CMP, (b) after 30 sec polishing and (c) after 10 min polishing.

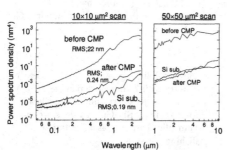

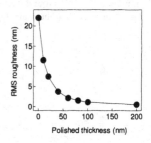

Figure 4. Power spectrum densities obtained from AFM measurements of the $Si_{0.7}Ge_{0.3}$ virtual substrate before and after CMP and a Si substrate.

Figure 5. RMS roughness values of the $Si_{0.7}Ge_{0.3}$ virtual substrate as a function of polished thickness.

It was revealed, however, that the roughness slightly increases during the post-cleaning due to etching effects of not only NH_4OH that etches Si but also H_2O_2 that solves pure Ge. The cleaning temperature dependence of the RMS roughness is shown in Fig. 6. The RMS roughness before CMP (as-grown surface roughness) was very large and different among different SiGe virtual substrates, but it is seen that all samples are successfully planarized by CMP and that the RMS roughness is several nm. However, the RMS roughness is seen to increase as the cleaning temperature increases for all samples, indicating that the etching effects of the cleaning reagents increases the surfaces roughness. The origin of the roughening is considered to be inhomogeneous in-plane strain-field distribution coming from nonuniform misfit dislocation distribution in the SiGe layer [6,7]. The observed variation of the roughness among the samples reflects the different dislocation structure and/or crystal quality depending on the SiGe growth methods and temperatures. At 20 °C, the RMS roughness is 0.4 ~ 0.5 nm for all samples and no variation is seen, which indicates that the roughness is not enhanced by the etching effect at relatively low temperatures. It was also found that the roughness strongly depended on the ratio of NH_4OH to H_2O_2, and the ratio was optimized so that the roughening should be minimized without reducing cleaning efficiency [15].

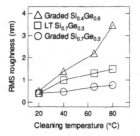

Figure 6. NH_4OH + H_2O_2 cleaning temperature dependence of the RMS roughness for several types of SiGe virtual substrates.

Regrowth on the planarized SiGe virtual substrates was performed and the quality of the regrown film, which is very important for application to channel structures, was evaluated by cross-sectional transmission electron microscope (XTEM). The regrowth procedure is as follows. After CMP and post-cleaning of the $Si_{0.7}Ge_{0.3}$ virtual substrate, H_2SO_4 + H_2O_2 (1 : 1) cleaning was carried out for 10 minutes to remove organic and metal contaminations, followed by the surface H-termination by a HF dip. Subsequently, the wafer was loaded into a MBE chamber and heated at 700 °C for 10 min. The temperature was then reduced to growth temperature of 600 °C, and a 100-nm-thick $Si_{0.7}Ge_{0.3}$ and a 25-nm-thick strained-Si layers were grown successively.

Figure 7. Cross- sectional TEM images of strained-Si / $Si_{0.7}Ge_{0.3}$ layers regrown on the planarized $Si_{0.7}Ge_{0.3}$ virtual substrate.

Figure 7 shows XTEM image of the regrown layers. It is found that the regrowth interface is invisible and the strained-Si layer without defects is formed. AFM observation also confirmed the flat surface equal to the as-planarized surface. Moreover, clear photoluminescence was obtained from the Si/Ge quantum well structures grown on the planarized SiGe[14]. These results demonstrate that the high-quality Si/Ge structures can be formed on the planarized and properly cleaned SiGe virtual substrates.

Application to strained Si and strained Ge channels

Strained-Si and strained-Ge channel modulation-doped (MOD) structures were formed on the SiGe virtual substrates planarized by CMP and transport properties were investigated. The fabricated structures were illustrated in Fig.8. For strained Si and Ge channels, the Ge concentrations of SiGe virtual substrates were chosen to be 25-30% and 65%, providing 1.0-1.3% tensile and 1.5% compressive strains into the Si and Ge, respectively. After the same cleaning procedure as described above, the MOD structures were grown as follows.

For the strained Si MOD structure (Fig. 8(a)), a $Si_{1-x}Ge_x$ buffer layer (100 nm), a strained Si channel layer (15 nm) and a $Si_{1-x}Ge_x$ spacer layer (20 nm) were successively grown with GSMBE at 660 °C. After the growth, the wafer was transferred to the SSMBE chamber and an Sb delta doping layer (3×10^{12} cm^{-2}), an undoped $Si_{1-x}Ge_x$ layer (40 nm) and a Si cap layer (5 nm) were successively grown at 300 °C so that segregation of Sb was prevented. For recovery of the crystal quality, the wafer was annealed at 700 °C for 10 min after the growth.

For the strained Ge MOD structure (Fig. 8(b)), $Si_{0.35}Ge_{0.65}$ buffer layer (50 nm) was grown with SSMBE at 500°C, and the substrate temperature was lowered down to 300°C and strained-Ge channel, $Si_{0.35}Ge_{0.65}$ spacer (10 nm), B-doped $Si_{0.35}Ge_{0.65}$ (10 nm), $Si_{0.35}Ge_{0.65}$ capping (40 nm), and Si capping (3 nm) layers were successively grown. Ge channel thickness was varied from 7.5 to 20 nm. As references, the same structures were fabricated on the unplanarized buffer layers for both Si and Ge channels.

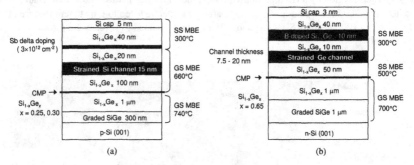

Figure 8. Schematics of (a) strained-Si and (b) strained-Ge modulation-doped structures fabricated on the planarized $Si_{1-x}Ge_x$ virtual substrates.

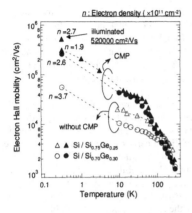

Figure 9. Temperature dependence of the electron Hall mobility for strained-Si modulation doped structures formed on $Si_{0.75}Ge_{0.25}$ and $Si_{0.70}Ge_{0.30}$ virtual substrates with and without CMP. Electron densities (n) are indicated for plots at 0.3 K.

Temperature dependence of electron Hall mobility is shown in Fig. 9 for strained-Si MOD structures formed on $Si_{0.75}Ge_{0.25}$ and $Si_{0.70}Ge_{0.30}$ virtual substrates with and without CMP. It is clearly found that the mobilities of the CMP samples are much higher than those of the samples without CMP at low temperatures. The highest mobility of 520,000 cm^2/Vs is obtained at 0.3 K with illumination for the CMP sample formed on the $Si_{0.75}Ge_{0.25}$ virtual substrate. The enhancement factor via CMP reaches as high as 4. This result clearly indicates that the main scattering mechanism is assigned to the roughness of the SiGe buffer and the overgrown channel layers and that the scattering related with the roughness can be drastically eliminated by planarization of the SiGe buffer layer. Above 100 K, however, the mobility difference between the samples with and without CMP becomes negligible, which suggests that the parallel conduction and/or phonon scattering make the effect of the surface smoothness invisible.

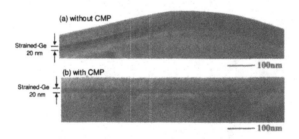

Figure 10. Cross-sectional TEM images of strained-Ge modulation-doped structures formed on $Si_{0.35}Ge_{0.65}$ virtual substrates (a) without and (b) with CMP.

Figures 10 show XTEM images of the 20-nm-thick Ge channel samples with and without CMP. Although 20 nm is beyond the critical thickness, high-quality strained-Ge channel layers are formed with abrupt heterointerfaces and channel thickness is laterally uniform for both samples. A considerable difference is the long-ranged roughness that the buffer layer initially has. For the unplanarized sample, the Ge channel is found to be largely bended along the undulation with a large height and a long wavelength. As a result, the strained-Ge channel layer in part deviates from the (001) plane by an angle of 20° at maximum. On the other hand, the channel of the CMP sample is very flat with no deviation from the (001) plane.

Temperature dependences of hole Hall mobility and density are shown in Fig. 11 for 12.5-nm-thick Ge channel samples with and without CMP. In the entire temperature range, the mobility of the CMP sample is higher than that of the sample without CMP, while hole densities are almost the same. This strongly indicates that significant carrier scattering is caused by the long-ranged roughness of the unplanarized buffer layer and is removed by CMP. Figures 12 show the channel thickness (T_{ch}) dependence of the hole Hall mobility at 10 K and 300 K for the samples with and without CMP. At both temperatures, it is clearly found that the mobility drastically decreases with decreasing channel thickness for the sample without CMP, while no reduction is seen and high mobility is maintained for the CMP sample. The mobility enhancement by CMP, as shown in insets of Fig. 12, reaches 8 and 1.8 times at 10 K and 300 K, respectively, at T_{ch} of 7.5 nm.

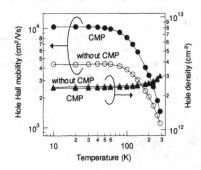

Figure 11. Temperature dependences of hole Hall mobility and density for 12.5-nm-thick Ge channel MOD structures with and without CMP.

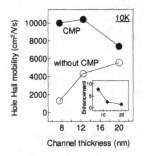 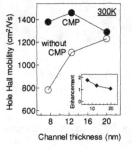

Figure 12. Channel thickness (T_{ch}) dependence of the hole Hall mobility at 10 K (left) and 300 K (right) for the strained-Ge MOD structures with and without CMP.

The strong T_{ch} dependence observed in the samples without CMP may be caused by hetero-interface roughness. When T_{ch} decreases, the hole wavefunction amplitude at the interface increases and consequently the interface roughness scattering increases. The interface roughness which dominates the scattering is not thought to be the long-ranged roughness visible in the TEM (Fig. 10), because it has much larger length scale than the Fermi wavelength of holes and the channel thickness. Therefore, it is reasonably speculated that the main origin of the scattering in the sample without CMP is not the long-ranged roughness but atomic-scale micro-roughness, which may exist at the tilted interfaces observed in the unplanarized sample. The fact that the CMP sample does not have such T_{ch} dependence and high mobility is obtained even in the thin channel suggests that the micro-roughness is significantly removed together with the long-ranged roughness by CMP.

To enhance mobility further, one should reduce the Ge composition of the SiGe virtual substrate and increase the strain [18], which requires thinner channels due to the decrease in the critical thickness. Thermal treatment during device fabrication also requires thinner channels to suppress the misfit dislocation generation. Therefore, since the Ge channel should be as thin as possible, the importance of the SiGe heterointerface properties largely increases for the device performances, and the smoothness of the interface becomes more critical. In this sense, the planarization of SiGe buffer layers will become essential for device applications.

SUMMARY

We developed chemical mechanical planarization (CMP) technique for the purpose of creating ultrasmooth SiGe virtual substrates. Adopting the similar process as Si CMP, we demonstrated that an initial roughness of larger than 10 nm was reduced considerably down to less than 1 nm. Post-CMP cleaning is an additional issue of great importance for regrowth of device channel structures. We found that the roughness increased during post-CMP cleaning process due to laterally nonuniform etching rate of the SiGe surface. To overcome this, we optimized the cleaning procedure, especially cleaning solution, and successfully obtained very smooth surface with RMS roughness of less than 0.4 nm, the lowest value ever obtained for SiGe surfaces.

The planarized SiGe virtual substrate was applied to strained channel structures. Strained Si modulation doped structure was fabricated on the planarized SiGe virtual substrate with a Ge concentration of 25-30 %. The electron Hall mobility obtained from the structure with CMP was 4 times higher than the reference structure without CMP, which is a clear evidence that the roughness-related carrier scattering can be well suppressed by the planarization. Strained Ge channel modulation doped structure was also fabricated on the SiGe virtual substrate with much higher Ge concentration (65 %). Although the roughness was much larger than 10 nm due to the high Ge concentration, the surface smoothness less than 1 nm was obtained by CMP. As a result, hole mobility enhancement factor over the reference sample without CMP was found to reach as high as 8 at low temperature and 1.8 at room temperature. These results clearly indicate that CMP is very promising technology for high performance strained Si/Ge based CMOS circuit.

REFERENCES

1. S. E. Thompson, M. Armstrong, C. Auth, S. Cea, R. Chau, G. Glass, T. Hoffman, J. Klaus, Z. Ma, B. Mcintyre, A. Murthy, B. Obradovic, L. Shifren, S. Sivakumar, S. Tyagi, T. Ghani, K. Mistry, M. Bohr, and Y. El-Mansy, *IEEE Electron Device Lett.* **25**, 191 (2004).
2. E. A. Fitzgerald, D. G. Ast, P. D. Kirchner, G. D. Pettit, and J. M. Woodall, *J. Appl. Phys.* **63**, 693 (1988).
3. E. A. Fitzgerald, Y. -H. Xie, M. L. Green, D. Brasen, A. R. Kortan, J. Michel, Y. -J. Mii, and B. E. Weir, *Appl. Phys. Lett.* **59**, 811 (1991).
4. H. Chen, L. W. Guo, Q. Cui, Q. Hu, Q. Huang and J. M. Zhou, *J. Appl. Phys.* **79**, 1167 (1996).
5. K. K. Linder, F. C. Zhang, J.-S. Rieh, P. Bhattacharya, and D. Houghton, *Appl. Phys. Lett.* **70**, 3224 (1997).
6. K. Sawano, S. Koh, Y. Shiraki, N. Usami, and K. Nakagawa, *Appl. Phys. Lett.* **83**, 4339 (2003).
7. Kentarou Sawano, Noritaka Usami, Keisuke Arimoto, Kiyokazu Nakagawa, and Yasuhiro Shiraki, *Jpn. J. Appl. Phys.* **44**, 8445-8447 (2005).
8. M. T. Currie, S. B. Samavedam, T. A. Langdo, C. W. Leitz, and E. A. Fitzgerald, *Appl. Phys. Lett.* **72**, 1718 (1998).
9. M. T. Currie, C. W. Leitz, T. A. Langdo, G. Taraschi, E. A. Fitzgerald, and D. A. Antoniadis, *J. Vac. Sci. Technol.* **B19**, 2268 (2001).
10. N. Sugii, *J. Appl. Phys.* **89**, 6459 (2001).
11. N.Sugii, D. Hisamoto, K. Washio, N. Yokoyama, and S. Kimura, *Tech. Dig. -Int. Electron Devices Meet.* **01** 737 (2001).
12. Z. –Y. Cheng, M. T. Currie, C. W. Leitz, G. Taraschi, E. A. Fitzgerald, J. L. Hoyt, and D. A. Antoniadas, *IEEE Electron Device Lett.* **22**, 321 (2001).
13. L. J. Huang, J. O. Chu, D. F. Canaperi, C. P. D'Emic, R. M. Anderson, S. J. Koester, and H. –S. Philip Wong, *Appl. Phys. Lett.* **78**, 1267 (2001).
14. K. Sawano, K. Kawaguchi, T. Ueno, S. Koh, K. Nakagawa, and Y. Shiraki, *Materials Science & Engineering* **B89**, 406 (2002).
15. K. Sawano, K. Kawaguchi, S. Koh, Y. Shiraki, Y. Hirose, T. Hattori, and K. Nakagawa, *J. Electrochem. Soc.* **150** G376 (2003).
16. K. Sawano, Y. Hirose, S. Koh, K. Nakagawa, T. Hattori and Y. Shiraki, *Appl. Phys. Lett.* **82** 412 (2003).
17. Kentarou Sawano, Yasuhiro Abe, Hikaru Satoh, Kiyokazu Nakagawa, and Yasuhiro Shiraki, *Jpn. J. Appl. Phys.* **44**, L1320-1322 (2005).
18. Kentarou Sawano, Yasuhiro Abe, Hikaru Satoh, Kiyokazu Nakagawa, and Yasuhiro Shiraki, *Appl. Phys. Lett.* **87**, 192102 (2005).

Mater. Res. Soc. Symp. Proc. Vol. 1157 © 2009 Materials Research Society   1157-E13-03

Xun Gu [1], Takenao Nemoto [2], Yasa Sampurno [3,4], Jiang Cheng [3], Sian Nie Theng [3,4], Ara Philipossian [3,4], Yun Zhuang [3,4], Akinobu Teramoto [2], Takashi Ito [1], Shigetoshi Sugawa [1], and Tadahiro Ohmi [2,5]

[1] Graduate School of Engineering, Tohoku University, Aza-Aoba 6-6-5, Aramaki, Aoba-Ku, Sendai, Miyagi 980- 8579 Japan
[2] New Industry Creation Hatchery Center, Tohoku University, Sendai, Miyagi 980-8579 Japan
[3] University of Arizona, 1133 James E. Rogers Way, Tucson, AZ 85721 USA
[4] Araca, Inc., 2550 East River Road, Suite 12204, Tucson, AZ 85718 USA
[5] World Premier International Research Center, Tohoku University, Sendai, Miyagi 980-8579 Japan

ABSTRACT

A novel end-point detection method based on a combination of shear force and its spectral amplitude was proposed for barrier metal polishing on copper damascene structures. Under some polishing conditions, the shear force changed significantly with polished substrate. On the other hand, the change in shear force was insignificant under certain polishing conditions. Therefore, a complementary end-point detection method by monitoring oscillation frequency of shear force was proposed. It was found that the shear force fluctuated in unique frequencies depending on polished substrates. Using Fast Fourier Transformation, the shear force data was converted from time domain to frequency domain. The amplitude of spectral frequencies corresponding to the rotational rate of wafer carrier and platen was monitored. Significant frequency amplitude changes were observed before, during and after the polished layer transition from barrier film to silicon dioxide film. The results indicated that a combination of shear force and its spectral amplitude analyses provided effective end-point detection for barrier CMP process.

INTRODUCTION

As technology node progresses, chemical mechanical planarization (CMP) becomes a crucial method in integrated circuit fabrication. It was reported that CMP reduces the total process steps involved in the device fabrication by as much as 25 percent.[1] During CMP, a rotating wafer is pressed against a rotating polishing pad with a certain pressure. A real-time end-point detection technique is required to ensure that the wafer is planarized properly and achieve constant yield.[2]

As part of the copper damascene process, Ta/TaN is deposited on patterned inter-layer dielectric (ILD) before copper deposition. The thin Ta/TaN layer acts as a liner to prevent copper from diffusing and also helps copper adhesion. During copper CMP, to remove the overburden Cu, Ta/TaN becomes a stopping layer to prevent over-polishing of copper lines. The optimal process time for Ta/TaN polishing is critical, especially when the Ta/TaN to ILD removal rate selectivity is low.[3] While the end-point detection (EPD) methods based on motor current, optical detection, and shear force measurement are widely used in commercial CMP equipment, there are many difficulties and issues for these EPD methods.[4] For example, the shear force values have been reported to be affected by polishing kinematic parameters, slurry concentration, and slurry chemical properties.[5] In this study, in addition to real-time shear force measurement, a complimentary EPD method is proposed to monitor shear force oscillation frequencies during barrier metal polishing. Fast Fourier Transformation is performed on the measured shear force data to analyze the amplitude of shear force oscillation frequencies under different polishing conditions.

EXPERIMENTAL

All polishing experiments were performed on an Araca APD-800 polisher and tribometer which is equipped with the unique ability to acquire shear force and down force in real time.[3] The force acquisition frequency was 1,000 Hz. Fast Fourier Transformation is performed to convert the measured force data from time domain to frequency domain and to illustrate the frequency amplitude distribution of shear force and down force.[6] 200mm blanket $Cu/TaN/SiO_2$ wafers and single damascene (SD) patterned wafers were polished. To prepare a SD patterned wafer, 30-nm TaN film as a barrier layer and 80-nm Cu film as a seed Cu layer were deposited by physical vapor deposition (ULVAC-ENTRON) without air break, followed by 700-nm electrodeposited Cu film, as shown in Fig. 1 (a). Prior to TaN polishing, the overburden copper layer was removed by copper CMP as shown in Fig. 1(b). Figure 1(c) shows the SD pattern structure after TaN polishing. Blanket $Cu/TaN/SiO_2$ wafers were prepared in the same procedures as the SD patterned wafer.

TaN polishing was performed on a Rohm and Haas embossed Politex REG pad at the pressure of 1.5 PSI (10,340 Pa). Hitachi Chemical HS-T815-B1 slurry with silica abrasive was used and the slurry flow rate was 300 ml/min. A 3M PB32A brush was used to condition the pad at 3 lb between wafer polishing. The brush rotated at 95 RPM and swept 10 times per minute. Wafer carrier rotational rate was varied from 23 to 60 rpm and platen rotational rate was varied from 25 to 63 rpm. The total polishing time for each wafer was 50 seconds.

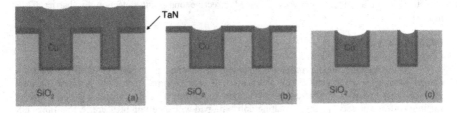

Figure 1. Illustration of the single damascene patterned wafer (a) before polishing, (b) after Cu polishing, and (c) after TaN polishing.

RESULTS AND DISCUSSION

Figure 2 (a) shows the shear force measurement during TaN polishing for a blanket $Cu/TaN/SiO_2$ wafer at the platen rotation of 25 rpm and wafer carrier rotation of 23 rpm. Based on the shear force value, there are three different polishing regimes as follows: 1) the bulk removal of TaN layer (i.e. less than 34 seconds); 2) the subsequent layer transition where both TaN and SiO_2 films were polished (i.e. between 34 and 43 seconds); 3) the over-polish regime (i.e. after 43 seconds). The shear force measurement effectively detected the end-point of TaN clearing at 43 s. Figure 2(b) shows the shear force measurement for a blanket $Cu/TaN/SiO_2$

wafer at the platen rotation of 63 rpm and wafer carrier rotation of 23 rpm. In this case, the shear force did not change significantly. This finding indicates that for an effective EPD, monitoring only the shear force value is not sufficient.

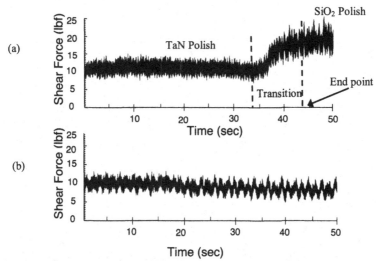

Figure 2. Shear force measurement of blanket Cu/TaN/SiO$_2$ wafer polishing at (a) platen rotation of 25 rpm and wafer carrier rotation of 23 rpm, and (b) platen rotation of 63 rpm and wafer carrier rotation of 23 rpm.

    The frequency spectra of the shear force before and after the layer transition at the platen rotation of 25 rpm and wafer carrier rotation of 23 rpm are shown in Fig. 3(a) and (b), respectively. There is a peak at 0.4 Hz that corresponds to the frequency of wafer carrier rotation and platen rotation. The amplitude of this particular peak increases by about 10 times after the transition to SiO$_2$ layer polishing. We define this frequency as a "characteristic frequency". It must be noted that there is a harmonic peak at 0.8 Hz as shown in Fig. 3(a) and (b).

    The frequency spectra of the shear force before and after the layer transition at the platen rotation of 63 rpm and wafer carrier rotation of 23 rpm are shown in Fig. 3(c) and (d), respectively. There are peaks that dominate at 0.4 and 1.0 Hz. The amplitude of these peaks increases by about 10 times after the transition to the SiO$_2$ layer polishing. Therefore, frequencies of 0.4 and 1.0 Hz are defined as the characteristic frequencies under this polishing condition. Frequency of 0.4 Hz corresponds to the wafer carrier rotation and frequency of 1.0 Hz corresponds to the platen rotation.

    By comparing Fig. 3(a) with Fig. 3(c) and Fig. 3(b) with Fig. 3(d), it shows that the spectral amplitude at the frequency of 0.4 Hz under the wafer carrier rotation of 23 RPM and platen rotation of 25 RPM is significantly higher than that under the wafer carrier rotation of 23 RPM and platen rotation of 63 RPM. As both the wafer carrier and platen rotate at about 0.4 Hz, they contribute to higher spectral amplitudes at 0.4 Hz.

Figure 4 shows the characteristic frequencies as a function of wafer carrier rotation with a fixed platen rotation of 25 rpm and platen rotation with a fixed wafer carrier rotation of 23 rpm. Those results indicate that wafer carrier and platen rotation dictates the characteristic frequencies of shear force oscillation.

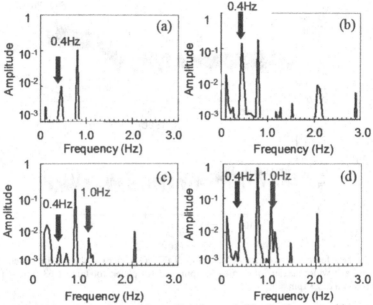

Figure 3. Spectral analysis of shear force (a) before and (b) after layer transition at platen rotation of 25 rpm and carrier rotation of 23 rpm and spectral analysis of shear force (c) before and (d) after layer transition at platen rotation of 63 rpm and carrier rotation of 23 rpm.

The amplitude at characteristic frequency of 0.4 Hz as a function of polish time at the wafer carrier rotation of 23 rpm is shown by diamond symbols (fitting result by the line) in Fig. 5. At the wafer carrier rotation of 23 rpm and platen rotation of 25 rpm, the spectral amplitude at 0.4 Hz starts to increase at approximately 34 s and reached a steady value at approximately 43 s. This indicates the end-point of TaN polishing also occurs at 43 s, which is consistent with the result from the shear force measurement shown in Fig. 2(a). At the wafer carrier rotation of 23 rpm and platen rotation of 63 rpm, the method based on the spectral analysis indicates that the end-point of TaN polishing occurs at approximately 43 s. In comparison, there was no significant change in the shear force value as shown in Fig. 2(b).

Figure 6 shows the SEM images of cross-section view of a patterned wafer after overpolishing for 15 s and 60 s. There is appreciable loss of SiO_2 and Cu after 60 s overpolishing compared with 15 s overpolishing. Resistance of Cu line (W/L=0.4 um/100 um) was measured by I-V measurement which the current of the Cu line was measured when the voltage is changed from 0 to 100 mV on Kelvin pattern. Cu line resistance increases from 11.6 Ω for 15 s overpolishing to 14.2 Ω at 60 s overpolishing. These results indicate that overpolishing causes changes in both patterned structure topography and electrical performance.

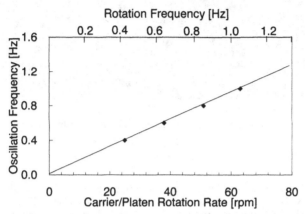

Figure 4. Characteristic frequency of shear force oscillation as a function of wafer carrier rotation with a fixed platen rotation of 25 rpm and platen rotation with a fixed wafer carrier rotation of 23 rpm.

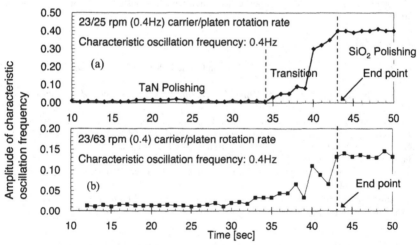

Figure 5. Amplitude of characteristic frequency of 0.4 Hz as a function of polishing time at the wafer carrier rotation of 23 rpm with (a) platen rotation of 25 rpm and (b) platen rotation of 63 rpm.

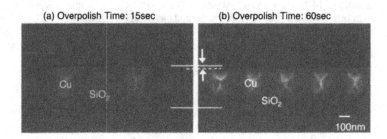

Figure 6. SEM cross section views on patterned wafer after overpolishing for (a) 15 s and (b) 60 s.

CONCLUSIONS

As the change in shear force was insignificant under certain polishing conditions, monitoring only the shear force value was not sufficient for an effective EPD. To improve the EPD method, shear force spectral analysis was performed through Fast Fourier Transformation and shear force oscillation frequency was studied. It was found that the shear force oscillation frequency amplitude changed significantly during the layer transition from TaN to SiO_2. The characteristic frequencies of shear force oscillation were dictated by wafer carrier and platen rotation. As a result, a novel end-point detection method based on a combination of shear force and its spectral amplitude was proposed for barrier metal polishing on copper damascene structures.

REFERENCES

1. P. B. Zantye, A. K. Sikder, N. Gulati, and Ashok Kumar, *Proceedings of CMP-MIC Conference* (2003).
2. H. Hocheng, and Y. Liang Huang, *IEEE Transaction on Semiconductor Manufacturing*, Vol. 17, No. 2, 2004.
3. Y. Sampurno, X. Gu, T. Nemoto, Y. Zhuang, A. Teramoto, T. Ohmi, and A. Philipossian, *Proceedings of International Conference on Planarization/CMP Technology*, p 132 (2008).
4. Gitis N and Mudhivarthi R, Tribology Issues in CMP, Semiconductor Fabtech, Henley Publishing Ltd, London, UK, 18th ed. 2003, p 125-128.
5. Y. Li, Microelectronic Applications of Chemical Mechanical Planarization, p 82, Wiley Interscience (2007).
6. Y. Sampurno, F. Sudargho, Y. Zhuang, M. Goldstein and A. Philipossian, *Thin Solid Films*, 516, 7667–7674 (2008).

AUTHOR INDEX

SUBJECT INDEX

Printed in the United States
by Baker & Taylor Publisher Services